Sustainability and the Design of
TRANSPORT
INTERCHANGES

Transport interchanges are central to the achievement of sustainable development. The ability to connect different modes of transportation, to stitch together flows of transport infrastructure and people in an attractive and coherent fashion, is critical to creating successful cities.

This book presents design principles for transport interchanges and offers analysis of best practice in the UK and abroad. The interchange is a new form of transport building which integrates into a single whole various modes of public transport, putting the passenger first (rather than the infrastructure). This book explores the phenomenon of the transport interchange with particular focus upon the need to achieve sustainable development through better design and management of transport infrastructure.

The aim is to demonstrate how this complex new building type integrates with the city on the one hand and with different types of transport on the other. In this integration, design in both plan and section is important as is urban and landscape design. Although several books exist on the design of individual transport building types (such as railway stations and airports), very few are available on the design of interchanges.

The idea of 'interchange' is increasingly relevant as town planners, engineers and architects address the question of sustainable development with its emphasis on energy efficiency, social cohesion, access for the elderly and urban regeneration.

Brian Edwards is Emeritus Professor of Architecture from ECA, Edinburgh, a corresponding member of the RIBA's Sustainable Future's Group and currently teaches and researches at the Royal Danish Academy of Fine Arts, School of Architecture, in Copenhagen.

Sustainability and the Design of

TRANSPORT

INTERCHANGES

Brian Edwards

Routledge
Taylor & Francis Group

LONDON AND NEW YORK

First published 2011
by Routledge
2 Park Square, Milton Park, Abingdon, Oxon OX14 4RN

Simultaneously published in the USA and Canada
by Routledge
270 Madison Avenue, New York, NY 10016

*Routledge is an imprint of the Taylor & Francis Group, an informa
business*

Typeset in Helvetica Neue by
Florence Production Ltd, Stoodleigh, Devon
Printed and bound in India by
Replika Press, Pvt. Ltd, Sonepat, Haryana

British Library Cataloguing in Publication Data
A catalogue record for this book is available from the British Library

Library of Congress Cataloging-in-Publication Data
Edwards, Brian, 1944–.
    Sustainability and the design of transport interchanges/
    Brian Edwards.
        p. cm.
    Includes bibliographical references and index.
    1. Terminals (Transportation) – Design and construction.
    2. Sustainable development. I. Title.
    TA1225.E39 2011
    711′.7 – dc22   201001899

ISBN13: 978–0–415–46449–9 (hbk)
ISBN13: 978–0–203–83965–2 (ebk)

Contents

Preface

Over the past few years the energy debate has shifted subtly from a focus on carbon emissions to wider questions of sustainable cities. This broadening has placed greater emphasis on transport energy consumption and as a consequence onto issues of infrastructure provision, town planning and public space. Energy and social welfare are closely linked in the arena of transportation. Mobility is a question of what is affordable in environmental and social cost terms.

This book raises a number of topics that have not received the attention they deserve in architectural or transport literature. A key subject addressed is the relationship between public transport, energy use, design and social sustainability. The gel that holds these strands together is the transport interchange – the place where people, energy use and infrastructure interface. Unfortunately, in the past the emphasis has been upon infrastructure and the role of engineering in creating the network of new facilities to serve the travelling public. This book takes a slightly different view. It argues that true social sustainability will only be achieved if people are put first, and their different cultural and social needs recognised. This broadening opens up the prospect of richer and more beautiful transport facilities, and one where the architect can work fruitfully alongside the engineer in creating them.

This book has benefited considerably from the authors experience of living in Scandinavia. Many of the examples are drawn from the Nordic tradition of investment in the social realm. This finds expression not just in the quality of design but the sense that transportation, urban design and regional planning are topics worthy of government intervention. The social democratic ideas in Sweden and Denmark in particular have led to the creation of many notable examples of transport interchanges in often modest towns which sit well away from normal gaze.

Architects have been involved in the design of most of the interchanges featured in this book. One recent trend is the importance attached to architectural rather than engineering design. Quality of life and quality of space go hand in hand in all buildings, but in transport architecture in particular. Moreover, architects do not design transport interchanges on their own – there are many partnerships and much technical collaboration. Central to this is the power of the design brief to put in place the values necessary to generate beautiful and sustainable transport architecture. The key to the 'brief' is the client which is one of the stakeholders that this book seeks to influence.

The shift from a fossil fuel to low carbon economy will mark the next few decades of the 21st century. Part of the re-structuring of our cities and countryside will be to make transport greener. This will entail new modes of transport, new technologies and new design approaches. This book is written to influence the latter.

Acknowledgements

The author wishes to thank Routledge for supporting the publication of this book. In particular, thanks go to Caroline Mallinder and more recently to Fran Ford for their support and encouragement over an especially long gestation period.

Thanks also go to the many architectural and engineering practices that have provided material for inclusion in this book. The author has been selective in the projects available for publication, believing that those featured in the book represent the best of recent design output. Mention should also be made of those practitioners who freely gave their time for interviews and subsequent discussion. In this regard special thanks go to Fiona Scott, Malene Freudendal-Pedersen, Christopher Blow, Sir Nicholas Grimshaw, Nille Juul-Sørensen and Leszeck Dobrovolsky.

The author also wishes to thank the many photographers who provided images often free of charge. As the book is in colour, the power of image is enormous in reinforcing key arguments.

Finally, the author wishes to thank the Royal Danish Academy of Fine Arts, School of Architecture, for its unfailing support for the research and teaching projects which form part of the basis of this book. Special mention and warm thanks go to Torben Dahl, the leader of the Academy's Institute 2, and Peter Henning Jørgensen and his students at Department 10 for testing and helping to refine many of the ideas outlined.

Brian Edwards
Copenhagen Sept 2010

Introduction

Transport interchanges are central to the achievement of sustainable development. The ability to connect different modes of transportation, to stitch together flows of transport infrastructure and people in an attractive and coherent fashion, is critical to creating successful communities. A combination of rising oil prices, growing anxiety over the consequences of global warming and a determination to address urban regeneration has resulted in greater attention being paid to public transport. In the switch from a largely private to public transport economy, the quality of design of transport facilities assumes great importance in encouraging acceptance of new forms of urban movement. The green market economy of the future is dependent upon achieving more sustainable forms of transportation, both of vehicles and of the wider systems that support the transport infrastructure. Here, transport architecture has an important part to play in helping to deliver sustainable development.

In addition to the drive for social sustainability, a number of trends reinforce political interest in public transport as governments search for new forms of low-carbon living. Mobility is crucial to the quality of life and, whereas the rich use their private cars, the poor use public transport. Wealthy societies travel more than poor ones and wealthy people more than poor people. However, as the cost of travel is falling, more people travel than before and they travel greater distances. Although transport is responsible for about 25 per cent of global carbon emissions, sustainable development requires mobility to be by energy-efficient public transport rather than inefficient private means. The twenty-first century will be one of car constraint, and of investment not in roads but in public transport infrastructure, and in integrated transport systems rather than individual modes of movement. In this shift in emphasis, the interchange emerges as an important, or at least rejuvenated, transport facility – one that is not justified by energy efficiency alone but by the search for wider social sustainability. In this the transport interchange is an emblem of sustainability acted out in light, space and mobility.

This book defines transport infrastructure as the network of routes, lines and links that provides for the movement of people and goods. The emphasis is upon public provision of such services. The transport interchange is the hub that permits the transfer from one mode of movement to another. As such, the interchange is a building or urban space that creates conditions whereby transport interconnection is achieved. Fundamentally, the interchange is about processing passengers between modes of mobility; the process is people-centred and consists of moving users from one transport source to another (Harbour, 2006). However, good transport architecture celebrates the mundane process of circulation and movement by creating uplifting spaces that, through their scale, volume and clarity, 'reduce stress and anxiety among the travelling public' (ibid.).

The characteristics of twenty-first-century mobility are as follows:

• People travel more often than ever.
• People travel further than ever.
• People in wealthy countries travel more than those in poor countries.
• Rich people travel more than poor people.
• Poor people use public transport more than rich people.
• Bus transport is the main mode of transport for poor people.
• Walking is the commonest form of mobility.
• Cycle use is increasing in cities at the expense of car use.
• The cost of travel is falling.
• Greenhouse gas emissions from transport are rising.
• Particle pollution from road traffic is rising.
• Air travel is the greatest polluter per travel mile.

This book is concerned with the design of transport interchanges at three main levels – urban, building and interior design – in order to address the principal drivers: infrastructure provision, community planning and people movement. The three main flows that stem (a) from the various modes of transport (train, bus, tram, ferry, bicycle, foot), (b) from the cross-currents of people movement (able and disabled) and (c) from the information displays and signs need to be integrated. However, they rarely are and the inability to coordinate the different interests results in frustration for the passenger and possible rejection of more environmentally acceptable methods of transport.

The interchange is a place of connection, of transfer and of social interaction. The structures that accommodate these travel movements need to serve interchange mobility well, both functionally and aesthetically. Unfortunately many examples, even some of those in this book, have ambivalent spaces where navigation is confused by the competing messages of commerce and transport provision. The design narratives may be clear but the realities of interchange life are far from perfect.

The interchange provides spaces for:

• interaction;
• transfer;
• integration;
• interconnection;
• intersection.

To achieve sustainable transportation, there must also be attention to ecological design of the different transport buildings involved. It is no good conserving fossil fuels in green transportation just to waste them through lack of interest in low-energy design of the interchange. There are surprisingly few examples of eco-design of railway and bus stations, and airport and ferry terminals. This means that the transport hub that connects them together has to make up for this indifference. What this book argues is for the need to connect green transport thinking with green design approaches at the interchange. After all, transport buildings are visited by millions every day and the examples of eco-design here, as well as the wider messages of environmental care, could be taken into the home and workplace.

The key to better transport design resides in better management and in better design. This book focuses on design in an attempt to change how transport managers procure new buildings and upgrade existing ones. The focus on design is also based upon the premise that consumers are influenced by the quality of design of facilities and transport-related spaces, not least those people with the ability to choose. Design also matters to those with less choice, such as individuals on low incomes or those people with limited sight or means of mobility.

One key argument in the book is the need to counter the sense of 'non-place' prevalent in the experience of transport buildings. In his book *Non-places: Introduction to an Anthropology of Supermodernity*, published in 1995, Marc Auge cites airports and bus stations as places that one remembers only in generic terms and hence without the normal specificness of place, time and culture associated with everyday architecture. Sadly, non-place is prevalent in many examples of recent transport architecture, particularly in some of the new Asian interchanges. The forces of internationalisation and of standard-isation do tend to unify the function and shape of architecture, especially within the transport realm. However, this book argues that attention to social needs, climatic and environmental forces, and cultural nuances, effectively temper the abstractions of globalisation and standardised engineering solutions. Such tempering holds the key to the provision in the twenty-first century of a generation of integrated and better connected transport buildings aimed at the post-fossil fuel age.

The transport interchange is:

- a gateway to mobility for many;
- a gateway to urban areas;
- a gateway to sustainable development.

Conceptually, the interchange exists between different gateways, movement patterns, land use functions and the web of connecting spaces. Interchange is both a process (for the passenger) and a place (for the architect and transport provider). It operates at different scales (feet and wheels), interfaces between landscape and urban elements and interacts in detail between buildings and their more intimate spaces. The interchange contains concourses, rooms and passageways that are colonised at variable levels of occupation and function in both programmed and unprogrammed ways. Hence, the architectural spaces created are mainly 'open' – they are waiting for people to give them life and colour as the passage of trains, buses, planes and ferries ebbs and flows.

A good interchange is rather more a framework for people than a fixed architectural monument, although in reality the best transport interchanges contain both fluid (over time) and fixed elements. However, like all transport buildings, they are transitory in nature both in themselves and in their impact upon the wider landscape. In cities the transport interchange has become a major catalyst of urban regeneration – a focus for commerce and the flow of ideas as well as the movement of people.

The interchange is also a place where the threads of urban life are joined together. In former times we referred to railway stations – places where the parallel lines of steel were dominant. Today we talk of transport interchanges – places as much for people as for trains, buses, planes and ferries. New transport interchanges are busy restructuring the growing urban centres of Asia and remodelling the established cities of Europe. They are places rather than spaces; they depend upon complex interdependencies, and they are full of contradictions. This is why as a building type they are so hard to design.

This book explores the phenomenon of the transport inter-change with particular focus on the need to achieve sustainable development through better design and management of transport infrastructure. It is concerned with people, with energy efficiency, with building and urban spaces, and with architectural delight. One theme running through the various chapters concerns the phenomenology of transfer space. Unless transport interchanges provide clarity of use and raise the spirit, they will not be able to achieve the switch from car use to public transport, no matter how well designed the new trains or buses may be.

PART 1

Concepts, ideas and evolution

Stations and terminals serve a transport network whose shape and technologies are in a state of perpetual evolution. The forces of change at a political, social and technological level form the basis of Part 1. The aim is to show how architectural design of the interchange is the result of wider forces within society. These forces vary over time and are subtly altered by the cultural and geographical nuances of different places. Part 1 seeks to explore the dialogue between design and engineering on the one hand and between the pressures of sustainability and mass mobility on the other. One recurring theme is that of the relationship between city growth and transportation, with consequences for the design and location of transport interchanges. Another is the dialogue between inherited patterns of transport use and the demands made by the low-carbon future.

   Part 1 is divided into two chapters. The first deals with mobility as a concept and questions of place and non-place, and of social inclusion and exclusion are addressed here. The second explores the relationship between transport, city form, urban image and technology. It debates the role of modernity and of public space in the context of transport architecture. A key element is that of inter-modality – how transport systems interconnect and how passengers transfer between types of transport. Rather than discuss the evolution of transport from single perspectives, Chapter 2 investigates the rise of the interchange as a late modernist phenomenon. The need to connect (and interconnect) is an expression of freedom in our segregated and functionally divided world.

Mobility and the interchange

CHAPTER

1

Mobility in developed countries accounts for around 60–100 minutes per day of a person's time irrespective of the mode of travel employed (walking, cycling, bus, metro, train). In spite of travel restrictions such as car-free zones, the distances travelled increase year by year. To a degree also, the time spent on travel increases annually but, as public transport becomes more efficient and interchanges more effective, the travel distance available within a set period of time expands. Hence suburbs and new towns are built ever further away from traditional urban centres. However, if public transport can take people further within a specified timeframe, this is not true of private transport, which suffers from two inefficiencies – congestion and regulation. The balance of power has switched in most of the world (though not the USA) from private modes of travel to public ones. The car is seen as increasingly un-environmental and also antisocial, and hence access for the car in urban areas in gradually being restricted. So in spite of greater affluence and more flexible lifestyles, the shift from private cars to public buses, trains and trams is a characteristic of the early years of the twenty-first century.

However, the change in travel mode is not universal, neither is it consistent across cultures and age groups. The age group of 25–35 is particularly wedded to the car. Here it is a question of lifestyle choice, of convenience – particularly for parents whose children need to attend school in low-density suburbs, and of image. The wealthy can afford to drive, the poor cannot. Even when there are incentives to use buses, such as free bus travel and prioritised bus lanes through congested streets, many in this age group choose the security and one-upmanship of the car to what is perceived to be less safe and socially acceptable trains and buses. Although car restraint is essential in order to change habits, the image of public transport and its connectors (stations, bus stops etc.) is a major obstacle to social change. So, although we need legislation, travel is a cultural statement as well as a functional necessity. We are how we travel or, put another way, we assert our sense of worth by the means of transport we adopt. So to ride a bike to work is to make a cultural statement; to choose to walk rather than drive is to send a message to your neighbours. Conversely, to use a large four-wheel drive vehicle in the suburbs shows a disregard for both the social and physical environment.

Many people carry prejudices about the quality of public transport, such as the 'trains are always late' or the 'metro is dangerous at night' or 'stations are filthy' or 'bus stations are full

1.1 Travel is a cultural statement as well as a functional necessity.
Oslo Station. (Photo: Brian Edwards)

of homeless people' (Freudendal-Pedersen, 2009). One role of design is to combat these poor images by projecting an optimistic, more utopian vision for public transportation. It is also important that existing problems with stations and interchanges are addressed, such as graffiti, poor lighting, vandalised seats, inadequate travel information and much else. Upgrading image is primarily a question of upgrading management practices and refreshing design. Also it is worth noting that those who complain most about public transport are those who don't use it, and those who make policy decisions tend to drive rather than take the bus or train. As a result, there is a gap in vision between those who take public transport as a matter of course and those who rarely do and who legislate on our behalf. The latter rarely wait around for trains, ferries or buses in dark stations or on windswept piers and streets.

Mobility does not have to be based on cars and on fossil fuels. Electrical energy produced from wind, solar and nuclear installations powers high-speed trains across Europe, and diesel used for many of the buses in Latin America and Asia comes from biofuels. Walking and cycling depend solely on human muscle energy produced by metabolism with only limited fossil fuel consequences. Hence, the totality of the mobility debate touches upon questions of energy use, access, equality and health. Put another way, urban transportation is as much about sustainable development as it is about the popular debate surrounding the greening of cities and low-energy architectural design. The major means of getting around are walking, cycling, car, taxi, bus, tram, train and aircraft. In reality we use some or all these modes often in conjunction with a single journey from home to work. Combining modes of travel raises the question of interconnection – of moving

1.2 In the USA the return to train travel offers many environmental benefits. Amtrak Station, Oakland, California. (Photo: Brian Edwards)

smoothly and efficiently from one type of transport to another. Here, the transport interchange comes into its own. The interchange recognises the interdependency of modern urban life – the plurality and complexity that underpin the contemporary situation. However, functional separation of land uses and their supporting transport modes, which was the mantra of modernism, has left a great deal of inefficiency and redundancy in our cities.

New transport modes, hybrid technologies and the future

One characteristic of transport buildings is that they are always having to adapt to innovations in the mode of transportation. This is not new. Steam that powered trains was replaced by diesel and then by electric traction motors, with single-deck carriages replaced by double-decker layouts; buses became hybrid fuel driven and extended into train-like configurations with concertina carriage joins; ferry boats became fast hover and then hydrofoil craft; trams adopted elements of bus and train technologies to run on city streets without pollution or excessive speed. The latest trams run on rubber tyres and use dedicated bus lanes, thereby blending the genes of rail and buses to create cost-effective mass urban transportation. In Curitiba, Brazil, such vehicles carry two million passengers a day, making the kind of expensive and disruptive tram infrastructure provision found in Edinburgh redundant. The story is one of stations, ferry terminals, airports and interchanges adapting to survive under relentless technological innovation.

Generally, technological change in the transport industry is driven by three forces: the need to improve efficiency, especially

1.3 Fluid interconnection is a characteristic of the well-designed interchange. This often involves modification to existing patterns of movement as here at King's Cross Station in London, redesigned by John McAslan + Partners. (Courtesy of John McAslan + Partners)

in energy consumption and people moving capacity; the need to exploit new markets in mass people movement; and the need to improve customer comfort levels. In Darwinian terms, it is not the most intelligent or strongest infrastructure that has survived, but those elements of public transportation that have proved most flexible and willing to change. Hence the interchange has emerged in response to new social and economic dynamics, departing in the process from the thoroughbred form of the traditional railway station or airport. The modern interchange is a hybrid that accommodates new transportation technologies and seeks to provide expression for the interconnected nature of modern multicultural and pluralistic life. The ability to handle large numbers of transferring passengers, to process complex movement patterns while also producing buildings of quality, comfort and architectural distinction, is the challenge ahead.

Looking forward there are several innovations on the horizon that may alter the pattern of transport use and hence the interchange. The first concerns the development by the Canadian company Bombardier of trains powered by lightweight jet engines. Designed for long-distance journeys, the jet-powered train offers better speed and acceleration without the cost of overhead electric power lines (Schilperoord, 2006: 120–7). The jet train, particularly if energy efficiency can be improved and biofuels exploited, may threaten the high-speed dominance of the French TGV, German ICE and Japanese Shinkansen technologies. Potentially too it could challenge air transport for mid-distance travel, particularly in the USA and Canada, where train infrastructure is underdeveloped and buses have withdrawn from the mid-haul market. Magnetic levitation technology (MagLev) also offers potential since weight is reduced and power efficiency increased perhaps tenfold over wheeled transportation. MagLev also overcomes the problem of noise, which is one of the major restraints on high-speed trains, especially in urban or suburban areas. Innovation here is centred in China with Shanghai being the first planned

1.4 Driverless metro train in Copenhagen, 35 per cent fuelled by renewable energy from wind and waste, presents a green vision of mobility for the twenty-first century. (Photo: Brian Edwards)

practical application (ibid.). Other innovations include monorail urban transit systems (which are already functioning well in Vancouver and Las Vegas), fuel cell bus technologies and hybrid bus–light rail formats. These dual road/rail vehicles run on both metal tracks and rubber tyres, creating flexibility at the urban–suburban–rural interfaces. A similar hybrid, although at less cost, is the triple articulated diesel bus, which runs in Curitaba in Brazil. Powered by bio-diesel, these have dedicated lanes and fully enclosed bus stops with automatic barriers akin to those employed on tram systems. These bus/tram hybids are less expensive than rail systems but employ similar technologies and offer higher levels of passenger comfort than traditional buses.

In terms of aircraft there is research and development in the area of double-decker planes, dual jet–glider configurations and hybrid plane–train technologies utilising high-speed train traction for aircraft. One characteristic of all these developments is that they offer greater energy efficiency and improve the operational performance of public transport. Necessarily, the stations, terminals and airports that serve the infrastructure of mobility need to be able to adapt to these external forces.

Two trends are evident, therefore, in the growth of transport design. The first is the pursuit of speed and fuel efficiency personified by the new trains currently under development in Japan and Germany. The second is the merging of transport vehicle types in order to better cater for passenger demand and offer greater flexibility to transport operators. This trend is having great impact upon the interchange since hybrid vehicle types, such as train–bus formats, are leading to the merging of functions at stations. These new hybrid public transport vehicles require hybrid spaces where passengers can move freely between the new transportation modes. Fluid spaces and more fluid management are required to deal with the breaking down of transport boundaries at interchanges. What is a bus

1.5 This hybrid bus–tram in rural Zealand links villages in the countryside with high-speed rail and ferry services. (Photo: Brian Edwards)

and what is a train and what is a tram are concerns of the past – the future suggests hybridisation of transport vehicle types and the emergence of distinctive interchange buildings to serve them.

The traditional train will not disappear since it offers many benefits for medium- to long-distance travel. What will change are the urban and suburban transport configurations that provide the links to the faster networks. Here, new vehicle designs will make interchanging easier and a more pleasant experience. These new format vehicles will carry bicycles and wheelchairs; they will replace the private car as the glue that joins the transport system together. It is here that the architectural challenge exists – how to accommodate the new transport technologies of the low-carbon age in an exciting generation of new interchange buildings.

Space inequalities in the city

One of the greatest inefficiencies is in the area per person devoted to different modes of travel. The space between cars and people in a typical city street is in the order of 5 to 1. Mobility on foot has just over 20 per cent of road space compared with that given over to car users. Cyclists, where there are dedicated cycle lanes, may enjoy 25 per cent of the space set aside for drivers or about the same as that of pedestrians. Bearing in mind the extra space between road vehicles needed for safety, the inefficiencies rise even more and double the calculation. So a pedestrian moving on foot may have only 10 per cent of the surface area of road compared to that of a motorist, and a cyclist perhaps 20 per cent. The lack of democracy in public space between polluting vehicles and unpolluting foot or cycle

1.6 Car-dominated interchange at Aarhus Station in Jutland, Denmark. (Photo: Brian Edwards)

1.7 Attractive bus–rail interchange at Mölndal, near Gothenburg in Sweden, designed as a new urban landmark by Wingårdh Architects. (Photo: courtesy of Ulf Celander)

movement is matched equally by time sitting at traffic lights. The resulting inequalities find their way to the interchange, where roads form barriers to pedestrian access, and where buses have to wait behind parked cars.

Linked to the question of urban space allocation is that of national space distribution. Roads have about 20 times the land area of rail, yet journeys by road are two to three times more energy demanding (and hence polluting) as journeys by rail. Also, nationally, the area allocated per person for travel is roughly equal to that for housing (at about 50 square metres per person). As the rail network is roughly three times as energy efficient as road, national investment across Europe is strangely geared to this unsustainable mode of movement. It also

supports least of all the most energy-efficient transport mode, which is muscle power – either by foot or bicycle (Hegger *et al.* 2008: B2.67).

As a rule, decreasing the speed of journeys and increasing the volume of passengers per travel unit increases the energy efficiency per person. Hence, buses and urban metro systems are relatively benign since their speeds are low and their carrying capacity high. They also have the ability to run on electricity, which reduces urban air pollution. With diesel rail transport, speed is an issue since to increase the journey speed from 50 to 75 km per hour raises the energy consumption by 50 per cent. High-speed rail therefore runs on electricity rather than fossil fuels, and to make the most of the energy consumed

1.8 Vauxhall Cross Bus Interchange designed by Arup Associates provides a fine new landmark to sustainable transport in London. (Photo: Brian Edwards)

there must be effective integration with suburban rail and metro systems that have their own logistics in terms of energy efficiency, speed and carrying capacity. This is why the interchange matters more today than in the past – energy efficiency, accessibility and levels of utilisation form a virtuous circle around the transport hub.

Defining the transport interchange

There are two main definitions. The first deals with infrastructure: 'An interchange is a place or building where two or more modes of public transport interconnect'. The second deals with people: 'An interchange is a place or building for people to transfer between two or more modes of public transport'. Hence, combining the two one could define an interchange as 'a place or building where people transfer between interconnecting public transport services'. The key words are *connect*, *transfer*, *place* and *building*.

In this context, 'park and ride' facilities are not considered interchanges since only one form of public transport is involved in the transfer between modes (i.e. train and car).

The transport interchange can be a:

- railway interchange;
- airport interchange;
- ferry interchange;
- bus interchange.

Characteristics of transport interchanges

Transport buildings as a whole have to deal with rapid changes in capacity over a single day, and to cater for new mobility technologies every few years, while also staying responsive to new ticketing and information systems. Added to this, there are frequent changes in passenger-handling approaches, in security systems and customs controls. Yet in spite of all the

1.9 Clarifying existing arrangements while catering for new transport modes is the challenge faced at older transport interchanges, as here at London's King's Cross. This design is by John McAslan + Partners. (Courtesy of John McAslan + Partners)

external changes, transport buildings are expected to be architecturally distinctive and offer high levels of passenger comfort. They are symbolic and physical gateways to travel and arguably the most public of our architectural typologies (Harbour, 2006). However, although they often occupy cramped inner-city locations, transport buildings are expected to contribute to the public realm. The architect is therefore both civic designer and building maker – there is the need to shape both interior and exterior worlds simultaneously. With the creation of new transport interchanges, there is the opportunity to signal renewal, modernity and sustainability. With existing buildings the challenge is often that of overcoming inherited patterns of disconnection and poor design.

Many transport interchanges from the past are poorly linked to pedestrian routes and surrounding community or business areas. Poor connection, disrupted sight lines of key

facilities, hidden entrances and dangerous underpasses or windy bridges make up the landscape of typical public transport provision. Busy roads often form movement barriers while shops and cafés impede routes inside and outside stations, and the awareness of other modes of transportation is rarely considered by infrastructure providers. The modern transport interchange puts the pedestrian first. Movement on foot is given priority over movement by wheel; two narrow wheels have priority over four wide ones; the ability to perceive other modes of public transport, to read directional signs, and to walk through buildings with natural light, is essential to the enjoyment of travel. Bearing in mind that about half of all journeys start and end on foot, the walker is generally poorly provided for in cities and at transport facilities in particular.

Part of the problem lies in the lack of integrated transport planning. Transport plans usually have policies for trains, buses,

1.10 Woolwich Arsenal Station in south London links well into the web of community facilities. At night it glows as a beacon of social sustainability. (Photo: Brian Edwards)

airports, ferry provision and cycle provision but little in the way of concrete policies for their connection. Since transportation is often provided by different transport companies (sometimes operating competitively), there are few incentives for system interconnection. So airport companies build airports, port authorities construct ferry terminals, bus companies bus stations etc. The users are rarely considered; instead the focus is on the provision of infrastructure. As a result the needs of the train come before the passenger and engineering comes before design. So architects are left with trying to create spaces for people after the big decisions have been made by others. Added to this, the lack of policies for transport in an integrated sustainable sense means that passengers have to navigate their way through the competitive and ill-considered left-over spaces between the infrastructure of tracks, bridges, roads and turning areas. The challenge today is to better connect existing

provision by designing for integration rather than separation. Although there are examples later in the book of new interchanges in countries such as China and Japan, the task in much of Europe and the USA is how to redesign existing provision to improve interconnection and modal transfer.

One problem with interchanges is how to allocate the total budget between the different participating agencies. This is particularly difficult where new transport systems are threaded into existing networks. To take a new metro line under an existing station or to extend a railway line to an airport involves investment by both the new infrastructure provider and the existing one. How these costs are divided can be contentious. Often also the conversion of a singular transport system to an interchange requires adjustments to the urban realm. Who pays for the new squares and external circulating areas necessary to provide access for a multitude of providers? A clear division

1.11 Changing patterns of mobility influence the design of transport interchanges. In Copenhagen new forms of bicycle transport are emerging as citizens abandon the car under pricing restrictions. (Photo: Brian Edwards)

Table 1.1 Transport modes

Type	Features	Limitations	Carrying capacity	Stop frequency
Walking	• Limit to distance • Fitness required • Highly flexible	• Not suitable for elderly or disabled • Limit to baggage carrying • Not suitable in highly polluted areas	• 1	• Not applicable
Cycling	• Fitness required • Flexible • Relatively quick in urban areas	• Not suitable for elderly or disabled • Bicycle storage required • Limit to baggage carrying	• 1–2	• Not applicable
Car	• Flexible in suburban and rural areas • Suits family life • Relatively fast	• Limitations in urban areas • Expensive • Contributes significantly to global warming	• 1–6	• Refuelling required • Every 300–500 km
Bus	• Relatively flexible • Uses existing street space • Relatively energy efficient	• Requires bus stops • Requires interchanges with rail and metro • Disabled access problems	• 20–100	• 200–400 metres
Metro	• Relatively inflexible • Can use existing street space	• Underground systems expensive • Disabled access problems • Requires interchanges with rail and bus	• 100–200	• 600–800 metres
Local rail	• Inflexible • Stations are community and social magnets	• Expensive to run and maintain • Requires density of population • Periodic interchanges needed	• 200–300	• 3–5 km
Regional rail	• Inflexible • Stations are economic magnets	• Expensive to run and maintain • Requires large well-spaced towns • Stations are interchanges	• 250–400	• 10–40 km
High-speed rail	• Inflexible • Stations are major economic magnets • Lines form cross-national networks	• Expensive to build • Expensive to run • Requires interchanges at every stop	• 500–700	• 50–100 km
Plane	• Flexible • Big environmental impacts • New business location	• Airports expensive to build • Fuel costs limit growth • Interchange needed	• 100–800	• 300–500 km

of costs and responsibility is needed at the outset to avoid difficulties later. Often the municipality takes on the coordinating role, assigning money itself to those elements that are seen as enhancing civic design or achieving better social integration.

Since many existing interchanges are in unattractive places on the edges of city centres, there are urban design as well as transport design issues to resolve. Large transport buildings are often marred by busy adjacent roads, have fences that act as barriers to movement, lack landscaping and are often surrounded by dilapidated warehouses or former industrial sites. Such areas are not pleasant or safe to use, neither are they obvious places to locate new businesses, housing and community provision. While the demand for travel connection grows, making interchanges an essential element of twenty-first-century urbanism means turning such places into desirable locations for investment.

The realities of many existing transport interchanges are:

• poor connections between different transport modes;
• poor legibility of routes and facilities;
• poor connection to urban hinterland;
• poorly coordinated transport information;
• frustrating interior and exterior spaces.

Many people have an interest in transport interchanges: local and central government, private developers, transport infra-structure contractors, local communities and the passengers themselves. Unfortunately the latter are poorly represented on planning boards, yet they are numerically the most important users. Professionally there are many skills involved, from civil engineering to town planning, architecture and project manage-ment. Unfortunately, in the UK at least, designers have rarely held sway with civil and highway engineers, who are employed directly by the rail and tram companies, or by those project managers who serve the major contractors. Architects are

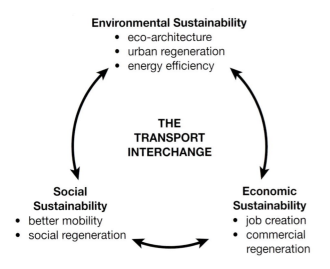

Environmental Sustainability
- eco-architecture
- urban regeneration
- energy efficiency

**THE
TRANSPORT
INTERCHANGE**

**Social
Sustainability**
- better mobility
- social regeneration

**Economic
Sustainability**
- job creation
- commercial regeneration

1.12 The three sustainable dimensions to the transport interchange. (Courtesy of Brian Edwards)

mainly employed after many of the critical decisions have been taken. This results in a lack of priority being afforded to stitching together the different transport modes, to connecting the interchange to the wider community, to ensuring integration of user and transport needs, and in putting beauty and delight into the equation.

Barriers to achieving effective transport interchange occur in:

- brief;
- design approach;
- facility provision;
- infrastructure dominance;
- cost;
- attitude of providers.

Mobility is a key measure of quality of life. Transport is also a significant user of fossil fuels (around 25 per cent of UK energy consumption and 18 per cent globally). Successful cities have effective transport interchanges that connect in an attractive fashion the worlds of work, living and leisure. As people switch from private cars to public transport in response to rising energy costs and greater urban travel constraints, the interchange assumes ever greater importance. The business case for better urban transportation is overwhelming. Companies want to be located where there are good transport connections, not just roads but rail, bus, cycleways and walkways. The social arguments in favour of public transport are also strong, particularly in an age of longer life expectancy and rising energy costs. Added to this, the growth in people migration following the erosion of national boundaries means the interchange is often the point of arrival in a city – a gateway to a new world of opportunity.

This book focuses primarily on the changing nature, typology and design of transport interchange buildings. A key concern is the changing function of interchanges in terms of their role as new urban magnets in the emerging imperative of 'sustainable development'. In this regard it is increasingly evident that the UK is behind much of Europe and emerging economies of

Asia in terms of the attention given to the design of transport interchanges and their attendant social and environmental role. The debate in planning and design circles has, however, shifted over the past decade from that of singular transport building types, such as stations and airports, to integrated inter-modal systems. The interchange, where two or more types of public transport infrastructure are brought together into a single building, is now becoming the norm particularly in metropolitan areas. As a consequence the interchange has an important contribution to make in achieving urban sustainability and good social welfare (Scott, 2003).

Achieving sustainable transportation requires:

- reducing the need to travel by land-use integration;
- making public transport as desirable and convenient as private transport;
- increasing urban densities to sustain public transport;
- changing modes of transport from cars to public transport, cycling and walking;
- improving the image of public transport;
- changing transport stops into interchanges;
- turning transport interchanges into business, community and sustainability hubs.

The basic methodology of transport studies – that of case study investigation, interviews with stakeholders, comparison of plans of contemporary examples and related urban layouts – has been followed in this investigation into the transport interchange. Since many transport buildings are easily visited and, of course, open to the public, there is much observational research. There is also a great deal of analysis of governmental policy on transport, energy efficiency, social inclusion and economic regeneration. Since local and central governments are invariably involved in funding major transport projects, their views form an important background to this study. The aim is to chart the design evolution of the interchange and to compare recent examples from around the world. Such analysis is useful to architects and engineers, allowing buildings to be studied as functional entities and changing typologies rather than as exercises in architectural fashion and style. However, this is not to deny the architectural drama of many recent transport interchange buildings or their innovative use of wide-span structural systems or clever environmental ones. In the analysis of case studies, the aim has been to reflect on how transport buildings mirror wider social, cultural and environmental change in society. Arguably, too, the interchange may not just reflect these changes but act as a catalyst for a greener future.

Transport interchanges should embody:

- smart infrastructure engineering;
- green transport technologies;
- green building technologies;
- sustainable urban design;
- sustainable communities.

Public transport, as the architect Lord Rogers pointed out in his Urban Task Force Report to the UK government in 2001, is one of the keys to developing more sustainable communities. Having mobility at reduced environmental cost and greater social inclusion benefits the city and the individual. To move by public means greatly reduces the carbon footprint of urban areas to the benefit of sustainability and hence society at large. But to facilitate this shift from private to public modes of transport we need good design, effective linkage between types of provision, and integration of urban design, communities and transport infrastructures. It is here that architects can play their part in shaping the future metropolis. After all, the primary concerns of any transport interchange are ease of access, shelter and legibility (Jones, 2006: 65) and these require the input of designers and engineers working together for the common good.

The transport interchange in the twenty-first century can be seen from three perspectives:

- as gateway to integrated travel modes;
- as gateway to towns and cities;
- as gateway to sustainable development.

How the interchange evolved

CHAPTER

2

Travel is a measure of civilisation and its speed or convenience a symbol of national prestige. The buildings that serve travel are also useful indicators of investment in the public realm for the common good, including that of sustainability. In many ways the history of transport architecture with its interest in space, construction technology, light and social provision is the story of modernism itself. Whether at the airport terminal, or railway or bus station, the infrastructure of public transportation has helped shape countless landscapes and cityscapes. Nowhere is this more evident than in the transport interchange with its concentration right in the heart of cities of competing social, economic, engineering and design forces.

In many ways transport is not only the movement of people and goods, it is also about the migration of ideas. Modernisation followed the paths of canals and then railways in the nineteenth century and airports in the twentieth. The stations and terminals that served the new transport corridors were places where ideas and new technologies were planted. Although the primary function was that of mobility, the secondary role of transport buildings concerned how those ideas were to benefit society at large. Hence, railways had a big impact on city location and urban form, and an equally big impact on economic structures and social frameworks. In this sense the stations (and today's airports) are cultural phenomena where people and ideas interface. This is why so many railway stations and early airports are now listed buildings – they are structures that embody cultural capital in their architectural design, engineering prowess and social impact.

The procurement of transport infrastructure has followed discernable cycles between public and private and between rail, road and air. The underlying explanation for the cycles is found in changing perceptions of who benefits from the transport investment and over what timescales (Marsay, 2008: 37). Around the world rail started by being mainly private investment and today is shored up by public subsidy, while road building was largely the result of public investment and is today expected to become more private in the form of toll roads. Airports, too, started as public investments and are now largely privatised. These cycles have implications for how the different types of transport infrastructure today are to be interconnected, who should pay and who benefits. Certainly it is not public subsidy alone that provides the explanation for the inherited pattern of infrastructure, since commercial interests are often involved. Few governments today are willing

2.1 Manchester Airport Bus–Rail Interchange. Adding interchange facilities at existing airports is a key feature of twenty-first-century transport investment. Notice the landmark qualities and high level of transparency – both intended to aid the needs of passengers. (Photo: Brian Edwards)

to bear the whole cost burden of smoother interconnection between public and often competing private transport provision. Since the benefits lie often in wider questions of economic regeneration and enhanced property values, there is an expectation of cross-subsidy and partnership funding. The question of social benefit and commercial gain suggests a balance between local and central government on the one hand and private venture capital on the other. This is the basis for the new St Pancras Interchange in London and the high-speed rail link to the Channel Tunnel and also the Fulton Street Transit Center in New York.

Historically, transport infrastructure was the domain of engineers, with little attention given to aesthetics or to people. Engineers determined the path of railways and canals, later roads and the location of airports and ferry terminals. It is infrastructure that gives the primary figure ground to landscape and the engineers who design the infrastructure determine also the shape of cities and the location of transport facilities. Their impact extends over a long period of time and the different forms of infrastructure, from transport systems to water supply,

provide one of the key agencies in achieving sustainable development. Only recently have architects become involved in infrastructure design beyond that of cosmetic effects. However, recent investment in public transportation systems, and in interchanges in particular, have opened up new opportunities for architects. Architects are now appointed alongside engineers in order to bring order and beauty to the interface between transport infrastructure and the city.

The interchange is a place of connection, public presence and movement. Basically, the interchange is the architectural expression of functionalism and movement acted out in interweaving patterns of interior space, light and technology. As with major railway stations, bus stations and airport terminals, the interchange is based primarily upon the geometries of movement, with large fluid spaces for social and economic exchange. Architectural design seeks to articulate the interface between the urban presence and transportation by giving form, legibility and imagery to these dramatic buildings. As the scale of provision expands, the architect's task of ordering space, structure, daylight and mobility in a coherent fashion becomes ever more difficult. Since transport buildings are both a gateway to travel and, in the opposite direction, a portal through which to experience cities and continents, the designer has the difficult job of turning the flows and counterflows of people into a memorable experience.

2.2 Shanghai South Station, designed by AREP. Many of the innovations in transport design and engineering are to be found in China. (Courtesy of AREP/T. Chapius)

Public transport buildings come in many types. There are railway stations (mainline, terminal and branch line), airport terminals (international, domestic and hub), bus stations (intercity and local), tram stops, ferry terminals (freight, drive-on, passenger and cruise) and interchanges (inner-city and city-edge) where inter-modal and hybrid types occur. The interchange is a place where two or more transport types interconnect with associated passenger facilities. This new hybrid began to emerge as the twentieth century closed. The new 'interchange' popular in France, Holland and Germany, as well as in China and Japan, is a structure where passengers move between rail, bus, plane or ferry in a single multi-decked building. The new interchange borrows much from earlier precedent, particularly the lofty day-lit public concourse, but differs in the complexity of travel types and hence the emphasis on way-marking.

All transport buildings share common characteristics of function and spatial configuration. The figure in plan includes a large gathering space, external connection via roads, bridges and pedestrian malls, interior linear spaces where travel is obtained (platforms, piers), an array of movement structures (lifts, escalators, staircases), and customer facilities of various kinds (ticket offices, shops, cafés etc.). This agglomeration of functions and facilities exists usually within a single all-embracing structure or a series of connected sub-structures arranged around a public space. Hence column, truss and beam are elements that provide a major part of the architectural experience. The need to provide orientation for often fatigued and disorientated passengers results in natural light and views being harnessed as a way-marking guide. Space (both internal and external), light (daylight and artificial) and structure (primary and secondary) form the principal architectural elements of typical interchanges.

Although transport buildings are primarily concerned with the efficient movement of people and associated means of transportation, the best examples also seek to provide dignity and hospitality for travellers. Dignity is reflected in the large columned spaces that constitute booking halls and the wide malls that channel passage; hospitality in the many shops and cafés that are constructed within transport buildings. Architecturally the invasion of commerce and retail into interchanges, railway stations and airports has often been at considerable aesthetic cost. The visual chaos that has followed has undermined the ordered architectural arrangement sought by the original designers. This is most marked in locations where the original conception embraced the coordinated design of finishes, furniture, lamp standards and advertising brands. As

2.3a Shanghai South Station, designed by AREP, re-orders the spaces of the city, altering investment and physical flows. (Courtesy of AREP/T. Chapius)

2.3b Generous well-lit spaces are essential to safety and security, as here at Lisbon Orient Station, designed by Santiago Calatrava. (Photo: courtesy of Torben Dahl)

the twentieth century closed, railway stations and airport terminals began to resemble anonymous shopping malls or nineteenth-century market halls. There are lessons here for the design of future transport interchanges.

When one examines earlier transport buildings it is evident that different architects have orchestrated the interaction of light, structure and volume in different ways. Although the 1851 Great Exhibition building in London (known as the Crystal Palace) inspired many nineteenth-century railway stations, notably London's King's Cross, Darlington Station in the north of England and Frankfurt Station in Germany, it was Eero Saarinen's creation at Dulles Airport in Virginia, USA – a sublimely elegant cage of ribs and glass tilted to the sky – that has inspired recent architects from Zaha Hadid to Santiago Calatrava. However, Charles Holden, who in various London railway stations set prisms of glass against brick cubes and cylinders in a fashion that landmarked his stations in the nondescript 1930s suburbs of north and west London, has had arguably a more enduring legacy, at least in terms of urban design. More recently, von Gerkan at Stuttgart Airport, Nicholas Grimshaw at Waterloo International, Norman Foster at Canary Wharf Station (both in London), and AREP at Shanghai Interchange have animated travel through a combination of large volumes bathed in natural light (legacy of Crystal Palace) coupled with structural boldness and sculptural expression (legacy of Saarinen). This combination marks the contribution to world architecture made by the current renaissance in transport design.

The most recent interchanges in Australia and China suggest the celebration of travel at a political and social level rather than the mere serving of passenger needs at a functional level, which was the case a generation earlier with stations in these continents. In this they have made connection with the earlier generation of transport buildings, which were the products of the first machine age. As such, they are 'celebrations of speed, industrial efficiency and the power of communication', as Lord Foster puts it (Foster, 2008: 11). These new transport buildings elevate interchanges to major urban magnets, creating new civic quarters around the concept of public rather than private transport.

In spite of the different manifestations of the architecture of transport in the twentieth century, the fundamentals of the

building type have changed little since nineteenth-century precedents. In *A History of Building Types* (1976), Nikolaus Pevsner devotes one of his 17 chapters to the evolution of railway stations. He starts by stating the obvious – that the building of stations presupposed the existence of railways. Hence airports also exist only after the introduction of movement by aircraft, bus stations after buses, garages after cars. The significant point is that new means of travel lead to the modification or evolution of transport building types. The structures that provide the means of access to transport systems respond to technological innovation in modes of transportation or changing patterns of transport planning, not to fashions in architecture. The current emergence of the significant new interchange type is a response to social, technological and environmental change whose roots lie in the concept of sustainable development.

Space and transport

The sense of lineage in transport buildings is important in understanding the resulting functional demands and the adoption of different kinds of layout by the early station designers. Railway stations were a particularly difficult architectural problem because no precedent existed in the nineteenth century for the type. Their architects evolved a form that later airport and bus station architects adopted. The key elements of the type (common to all early transport buildings) was a gathering space where tickets and information were obtained, a separate space where boarding occurs, and routes between the two. The first space is enclosed, often round and architecturally contained; the second is open and linear; the third is narrow and often marked by security or ticketing checks.

The geometry of space mirrors the pattern of activity involved in moving from a public to a semi-public realm, and in negotiating a change in speed facilitated by joining a train, plane or bus. Since these conditions are constant for all transportation building types, Pevsner is right in grouping stations, airports and ferry terminals together. However, the interchange adds considerable complication to the model. Now there are complex movement patterns often involving different physical levels

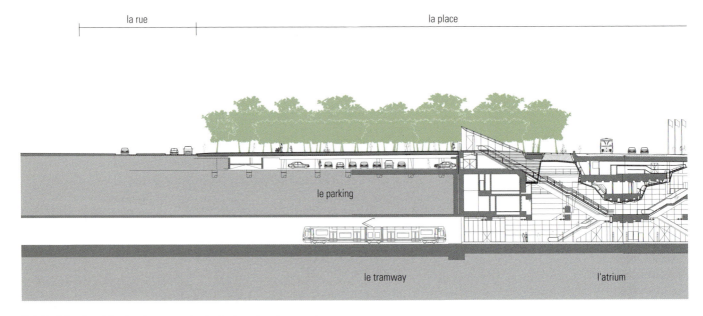

la rue | la place

le parking

le tramway | l'atrium

2.4 Variety of architectural spaces at a typical modern transport interchange. Strasbourg TGV Station, designed by AREP. (Courtesy of AREP)

and mobility at different speeds. There are often also complex ownership patterns that result from diverse franchising arrangements, sometimes involving competitive rather than cooperative action. Movement by different means usually involves travel on different levels, which entails people navigating steps, lifts, ramps, escalators and stairs. As the scale and speed of transport provision increases with the introduction of new modes of movement, the social and economic impacts expand. The interchange designer has much with which to grapple on three main spatial levels – urban, building and interior – and under three pressures – commercial, social and environmental.

Space is experienced on foot. The interchange is a set of connecting transport facilities where the gel that holds the system together is walking. Passengers invariably walk between bus, train, tram, taxi etc. Movement on foot is how the geometry of interchanges is experienced, how connection is achieved, and how places are memorised. As Le Corbusier (1927) put it, 'Architecture is experienced by eyes that see, a head that turns and legs that walk.' Walking is often the neglected dimension to interchange design, either within the transport structures themselves or around their edges. The neglect of journeys on foot in the spaces at the perimeter of stations results in much conflict between walking passengers and those in vehicles (which invariably have priority). Such contesting of sometimes chaotic or ill-defined urban space around interchanges leads to much frustration and sometimes even physical accidents. Planning for foot movement is the essential starting point in interchange design. After all, walking and cycling are the most energy efficient, healthy and sustainable modes of urban transportation (Tolley, 2003: 13) and need to better join up with the next best mode, namely public transport.

The twentieth century witnessed the expansion of transportation to the point where, in terms of passenger throughput, a facility such as King's Cross Station in London handled in a single year more than the whole population of England. The railway station, after a period of stagnation at mid century, was revitalised by innovations in high-speed traction (such as the French TGV system, the German ICE and Japanese Shinkansen). The 56 million passengers a year (in 2008) passing through King's Cross are taking advantage of innovation in high-speed rail technology that has helped transform King's Cross and its immediate hinterland from a large station to a major transport interchange. New connections by underground railway, new surface bus systems and, finally, the Eurostar link at adjacent St Pancras Station have conspired to turn a relatively run-down train terminus into a model of inter-modal efficiency. Working with the existing rail and bus infrastructure, King's Cross (like similar stations in Paris and Berlin) was gradually transformed into a hub, with four or more public transport systems woven around older facilities. However, as with many conversions from station to modern transport interchange, there is a loss of where the centre is located and a corresponding difficulty in gaining a sense of directional flow.

From relatively modest beginnings in the nineteenth century, the architecture of transportation emerged in the twentieth as a significant force in the evolution of new construction technologies and building types, leading eventually to full maturing in the form of the modern transport interchange. Such was their impact that transport buildings fashioned the spatial geography of cities and determined where the economic magnets were to be located. More recently, the transport

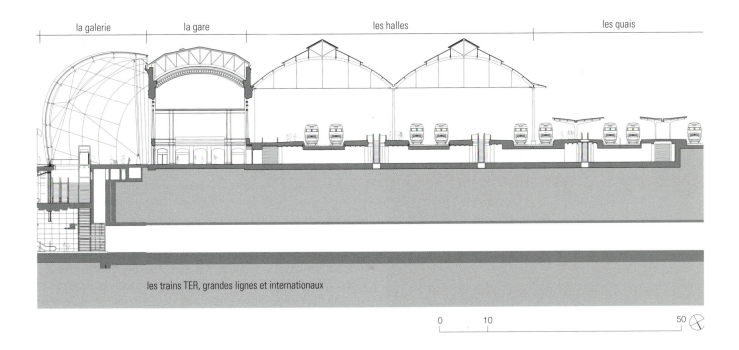

la galerie   la gare   les halles   les quais

les trains TER, grandes lignes et internationaux

0   10   50

interchange has begun to restructure the urban infrastructure on more sustainable lines.

New technologies and transport

The evolution of transport buildings in the 20th century provided the opportunity to test and develop new forms of construction technology that are seen increasingly in 21st century interchanges. Not only were large column-free spaces required for waiting rooms and booking halls, but roofs were often angled or cranked to allow for cross-ventilation. Many of the initiatives in concrete wide-span construction stemmed from continental Europe, where architect engineers such as Pier Luigi Nervi led the way. The tradition in Europe of the collaborative training of architects and engineers provided a shared knowledge base that encouraged technological and design invention. The post-war reconstruction of European cities led to innovation in railway station design, for example at Rome Terminal (1951) and Naples Central (1955), which influenced practice generally, not just at stations but at bus termini and airports. Similarly, the expansion of US airports in the 1950s, under pressure from the US government and airline companies such as TWA, led to the introduction of multi-storey terminals, which became commonplace throughout the world and have influenced modern interchange layouts. Such stations and terminals provided models of people movement, travel information and economic exploitation found in many transport interchanges today. They also signalled the importance of architectural design as a key element in user satisfaction and also to the branding of the cities and transport companies involved.

2.5 Transport architecture is a major force in shaping cities, as here in Tokyo. (Photo: Brian Edwards)

2.6 The great gathering spaces of the future are to be found within transport buildings, as here at Kansai Airport, designed by Renzo Piano. (Photo: Brian Edwards)

Size and longevity are also characteristics of transport architecture. Few building types are as large as transport buildings and not many examples of urban infrastructure survive as long as transport buildings. Bigness carries in its wake opportunities for architectural expression and long life suggests a responsibility for ensuring quality over time. Architects such as Eero Saarinen at TWA Terminal at JFK Airport and E.A. Riphagen at Schiphol Airport were early exponents of the ability to turn transport buildings into urban landmarks. It is a tradition that survives today in the work of Grimshaw, Foster + Partners, Renzo Piano Building Workshop, Zaha Hadid and Foreign Office Architects. Bigness, too, led to prefabricated construction and particularly the use of modular systems of high-tech steel and aluminium framing and toughened glass. One advantage of the industrialised system of building stations was the adoption of a language that unified the various parts and often extended to all the stations along a railway line. Prefabrication also allowed for replacing parts damaged by civil unrest, storms or vandalism.

Technological innovation has been a significant force in giving shape and identity to transport buildings. The need for large internal volumes lit by daylight and cooled by natural ventilation led to the adoption of intelligent wide-span structures. As a consequence, the modern interchange is invariably a tectonic essay in steel, glass and concrete. Walls, where they occur, tend not to reach to the roof. The result is the creation of subtly defined territories within bigger spaces, and when walls occur these are often given rounded shapes so as not to impede pedestrian movement. The presence of architectural structure is a recurring theme of this building type. Columns define routes and mark the edges of platforms, of concourse spaces, of ticket halls and of the building itself. In their turn, columns support roof trusses, which orientate the passenger inside and help give identity to the building within the external cityscape.

Space in earlier transport buildings came in three main forms: a gathering space represented at the railway station by the booking hall and at the airport terminal by the arrivals lounge; a linking concourse; and a transfer space characterised by the platform (station) and gate lounge (airport). As scale increased, so too did the complexity, but the distinction remained: three types of passenger space, one rounded and enclosed, one directional but contained, the other open, linear and in direct contact with the mode of transport. In addition, there was accommodation associated with transportation, such as offices, security areas and control points. The latter consisted of prominent structures overlooking the track in the case of the signal box, and the air traffic control tower at the end of the runway in the case of the airport. How these various elements are expressed or forged into a whole gave character or meaning to transport architecture. Increasingly, however, new types of accommodation, such as shops, which help to generate income, blur these functional distinctions and add confusion to passenger routes through transport buildings. So, today, the interchange is often a hybrid of earlier arrangements and consequently can make a less than coherent architectural statement, unless designed to express the fluidities of movement and scale of ambition inherent to the interchange.

The station and the city

The image of a transport building is one of large muscular structures set invariably in an area of open space. The space

2.7 The entrance to St Pancras Station, London. (Photo: Brian Edwards)

is needed for circulation and for visibility. The railway station is a gateway to the train, just as the airport terminal is to the plane. The concept of gateway leads to two major spaces: an outdoor one facing the city where external connections are made, and an indoor one focused around the concourse. This sense of gateway has influenced the architectural image of the building type, inspiring elegant internal malls and lofty concourses that lead the passenger to transport services. But in the opposite direction the station and airport are also gateways to cities and to continents. To arrive in this type of building is often to experience a whole new world, as at Delhi Central Station with its Hindu temple architectural language and at St Pancras in London with its Gothic style.

In the twenty-first century the supremacy of 'place' (rather than the abstractions of space) has gradually invaded the culture of modernity and eco-tourists are now as demanding of architecture as business travellers. Many examples around the world suggest that the new interchanges, whether at railway stations or airports, are becoming fashioned by the social, climatic and geographical nuances of the locality. In this sense, transport architecture is caught in the cross-currents of history, geography and the imperatives of climate change, and hence finds itself one of the new emblems of sustainablility.

Unlike many railway stations in the UK, those in Europe and the USA tend to have a more regular, planned relationship with surrounding areas. The station concourse and civic spaces are considered as one, with squares placed at station entrances and stations positioned along major road axes. For instance, the beaux arts-planned Union Station, Washington, designed by Daniel Burnham in 1903, has a large forecourt linked to the main booking hall, which leads axially to a grand passenger concourse. Penn Station in New York by McKim, Mead, and White, 1910, followed a similar plan, filling a whole city block with a high-level concourse above platforms. In both cases there was integration between modes of transport (mainly train, tram and bus), the station and the wider city. Similar planned relationships are found in Paris (Gare St Lazare, 1912), Frankfurt (Frankfurt on Main, 1915), Buenos Aires (Retiro Terminal, 1915) and Sydney (Central Station, 1910). Of the London termini it was only Waterloo, following the Memorial Entrance extension of 1922, which approached the planned grandeur of continental examples, although here later additions have undermined the

spatial clarity. These precedents are being re-examined today, as architects and town planners grapple with the complexities of modern interchanges and their city connections. As Lord Foster notes, big railway projects in the past created new civic quarters that were joined to older areas by public spaces employing 'good design in the civic realm' (Foster, 2008: 11). The civic realm has been a preoccupation of the architect Sir Terry Farrell, who notes that, in spite of their similarities, transport projects need to be grounded in the 'particularity of a city's culture and context' (Wong, 2008: 5).

Image and identity

By the early 1960s Britain had begun to realise that railways were both an asset and a problem. The overprovision of services that resulted from nineteenth-century free market competition led to inefficiency, which the Beeching Plan of 1962 sought to rectify. With the nation preferring the convenience of the car to outmoded public transport, Lord Beeching recommended the closure of 40 per cent of the UK rail network. His report made no mention of transport

2.8 Large interchanges need to address the language of the city, as here at Charing Cross, London, facing the River Thames. The design is by Terry Farrell and Partners. (Courtesy of Sir Terry Farrell)

interchanges and little of the need for social responsibility by train operators or the economic potential of rail infrastructure in urban regeneration. At the same time there was a growing awareness at British Rail of the benefit of good design to the effective management of a modern railway. The image-conscious 1960s led to a review of station environments, down to the level of graphic design and the uniforms of staff. New universal rules were introduced to coordinate the design and layout of stations, with particular attention paid to platforms and booking halls. With the legacy of the Festival of Britain still prevalent, the redesign of Chichester Station (architect N.G.T. Wikely, 1961) and the new Coventry Station (W.R. Headly, 1962) both symbolised the design renaissance. The image of the station and its graphics became even more important under the influence of the Design Council, which was established in 1964 partly in response to the perceived weakness of public transport architecture.

By the early 1970s railway stations and, to a greater degree, airport terminals, had become corridors of corporate and national identity. The values of a railway company, which historically extended from one end of the country to the other, were followed by airline companies that established a consistent image across national frontiers. The values were not only expressed in new forms of architecture, interior design and station graphics, but other factors such as computer use in signalling and programming became established along the transport corridors. Both railway and airline companies were in effect promoters of new technologies and of new ways of conducting business. Continents became developed on the back of the expansion of the railways in the nineteenth century and of airports in the twentieth. These linear avenues of development affected the way of life of ordinary people, and in this transport buildings were increasingly the emblem of modernity. However, the approach was driven by commerce, not wider social or environmental considerations. Provision was often competitive in spirit with little sense of an integrated transport network. Hence, Heathrow had little or no rail connection, and new stations such as that at Milton Keynes (to serve the new town of the same name) had no rail connection to regional airports and little bus provision.

The ambiguity between the competing agendas of architecture and engineering gave considerable charm to early transport buildings. Architects such as John Dobson in Newcastle sought a station façade that acknowledged the

2.9 The age of monumental railway architecture in Europe has left some notable civic structures to inspire today's designers. Atocha Station, Madrid. (Photo: Brian Edwards)

values of the city and a platform enclosure that addressed the train. Whereas the former was all pomp and circumstance, the latter was an experiment in wide-span construction, the daring use of wafer-thin glass and slender iron columns. Airports did not suffer the same sense of divided priority, mostly because airports exist in a world devoid of traditional urban values. The placeless airport required a response to the romance of flying, rather than to the physical landscape of the city. So early airports took their external imagery from the smooth, curved lines of the aircraft themselves, or from the wings of birds held in animated expectation.

The introduction of the TGV and ICE train networks in the 1980s established a high-tech corridor across Europe, periodically landmarked by new stations characterised by dynamic interactions of architecture and technology. In Africa, new airports also brought all the emblems of modernity in their wake – computers, cars, planes and mobile phones. In less

than a decade the savannah could be transformed into a successful business location with an international airport at its centre. The airport, like the station, is a technological transplant. Here one finds smart technologies, intelligent modern buildings and emerging business centres within a taxi ride radius. But these were single transport types, not multi-modal interchanges. Their functional singularity was typical of the age and left the question of transport interconnection to later generations. The lack of transport planning, certainly in the UK and USA, undermined the very concept of interchange between travel modes, and hence the associated social, economic and environmental benefits were not achieved. Design and technological image seemed to matter more than the serious question of moving goods and services at lower environmental cost.

The image of the station and of the airport became a tool used to market the company. The pointed arches or classical pediments of early stations, and the smooth uncluttered lines of modern airports, allowed one company to differentiate itself from competitors. Civic pride was expressed in transport

2.10a Platform canopy at Newcastle Station. (Photo: Brian Edwards)

2.10b Platform canopy at York Station. (Photo: Brian Edwards)

2.11 Karlsplatz Station, Vienna, designed by Otto Wagner, was one of several similar designs in central Vienna. (Photo: Brian Edwards)

architecture, first in major stations and later along the branch lines, just as national pride lurked behind the imagery of airports. The circular drum of Terminal 1 at Charles de Gaulle Airport (1964), designed by Paul Andreu, was a symbol both of Air France and of French fascination with monumental modernity.

Once a railway company had established its style, this was multiplied down the line. Hence, Thomas Telford's classicism found expression in countless stations along the Great Western and extended almost unbroken from London to Exeter. Likewise, on the Southern Region a cottage style influenced by the Arts and Crafts Movement stretched a consistent language across the home counties around London. Similar patterns of stylistic consistency are to be found in Vienna with the secessionist stations by Otto Wagner. Company image was reinforced by a palette of colours and patterns applied to the stations, the railway carriages, the interior upholstery and the uniforms of station staff. Even the hotels that existed alongside

the major stations (such as the Royal Station Hotel, York) shared in image building, since they were invariably owned by the same company. Hence, the interchanges of the day were simply enlarged versions of railway stations – not the multi-purpose, multi-imaged structures of today. But the question of image is less easy to address in modern interchanges. With different companies and stakeholders involved today, how can one avoid an architectural zoo of competing commercial graphic themes, house styles and architectural languages?

The airports and airline companies engage in a similar, though often less strident, form of travel advertising. Exterior architecture provides an opportunity for theming as well as the interiors themselves. Here, with airline companies such as Virgin and EasyJet, all is branded with the same house architecture, colours and graphics. It is generally the interior that matters, not the architecture of transport buildings in their entirety. The imposition of identity extends from the means of transport (planes), to the gate lounge, concourse area, staff uniforms and even the ticket graphics. Style, as such, is fashioned by the search for identity and commercial advantage in a business where place, time and distance are increasingly eroded. By way of contrast to the airline companies and retailers within the terminal malls, technology is seen as holding the key to wider architectural expression, resulting in an almost universal 'high-tech' language for late twentieth-century airports and their connecting railway stations.

The circular airport terminal was a feature of the 1930s and helped give identity to this new kind of transport building. One of the first of the type was at Gatwick, built in 1935, followed by Helsinki, 1938. Both contained radiating piers extending like fingers from a central round concourse. A ring of offices and bars separated the central public space from the customs hall, which formed an outer ring overlooking the runway. Although the plan evolved on the basis of an appropriate 'airport image' and of flexibility, it proved incapable of incremental growth and was subsequently abandoned in favour of rectangular terminals such as those found at Heathrow. Gatwick integrated airport and railway facilities (an idea revived in the UK in 1987 with Norman Foster's design of Stansted), where the circular shape allowed passengers to move smoothly down to the train using gentle curved ramps and staircases. The rotunda shape served as a useful icon for the emerging building type and has been employed for many recent transport interchanges (e.g. Shanghai South Interchange).

Land development and transport planning

The commercial ambition of a typical railway company extended beyond the immediate hinterland of the railway station. Land development was undertaken to exploit the opportunities that railways brought for economic and social expansion. The first phase of railway construction saw land around stations increasingly used for sidings and warehouses. By the end of the nineteenth century, this land was often redeveloped for hotels, offices and retail development. Major cities such as London, Leeds and Edinburgh had impressive hotels, which towered above the mainline stations, reaping profits for the railway company. The standardisation of architectural style between station and hotel was merely a reflection of a consistent approach to land development and ultimately of profit. However, there was little cooperation between the city councils of the day and the commercial developers. Land was parcelled up by railway companies for private sale, not given over to social purposes or for the inclusion of connecting transport services. So interchanges did not exist unless there was political pressure for wider transport connection (as with London Underground's Circle line, which joined up the major rail termini). Also, since large stations were seen as intrusive, they were often prevented from being constructed right in the heart of historic or commercial areas. Hence, London and Paris have their rings of termini, and York and Stuttgart Stations were built outside their city walls. Although many cities have expanded to embrace their nineteenth-century stations and ferry facilities, there is often the difficulty of transfer between transport types. One result was the construction of tram and metro services aimed not at suburban expansion but at inner-city interconnection.

Following the example of the railway companies, the developers of many metro systems saw their stations as local civic landmarks. This was the case in London and many other European capital cities. On London's District, Northern and Piccadilly lines, many of the new stations built in the 1920s and 1930s were given prominence in the street scene, often with squares in front for buses or marked by a clock tower. These stations, as at Morden Station in south London, were community centres in the broad sense, with shops and often council offices nearby. They stood as gateways to travel and as links between local civic ambition and national government transport policy.

Later in the century, airport authorities such as BAA played their part in providing land, whereby an airline company could build a terminal, hotel and car park as an integrated package, commercially and architecturally, often involving a transport interchange (e.g. Gatwick, 1970, and Stansted, 1987). Under this influence, the airport had matured by the end of the twentieth century into a new kind of transport-based city, with commerce at its centre. Most large airports contained the competing agendas of different airlines and hotel chains (each with its own design philosophy), as well as a multitude of retailers (SockShop, Starbucks, McDonald's etc.) within the buildings themselves and transport interchanges around their edges. The nineteenth-century station was a relatively measured and controlled environment compared to the modern-day airport, but the seeds of exploitation of the travelling public were sown in the railway terminals of London, Paris and New York.

By way of contrast, some local planning authorities have tried to influence the development of their cities by employing transport plans harnessing the entrepreneurial spirit of transport companies for the common good. A good example, typical of Scandinavian social planning, is the Finger Plan adopted for the Copenhagen region in 1947. Rather than allow for piecemeal growth, the plan proposed a number of dense corridors of housing, industrial and civic development along the lines of the expanded suburban rail network. The fingers of new urbanism were to help preserve wedges of countryside, which allowed ecological diversity to enter into the heart of the city. The resulting dense pattern of urbanism supported cycle and foot access to the stations, and later provided justification for inter-modal transport hubs at intersections between rail, metro and bus services. These hubs, though justified initially on social and transport grounds, have become the location for new business, leisure and educational development. Today, over 60 years later, the Copenhagen City Plan requires offices of over 1,500 square metres to be located within 600 metres of stations. The lesson here is one of cooperation between civic, transport and commercial developers, leading to a more sustainable city.

Basically, railway and bus stations and ferry and airport terminals exist where the needs of people to travel and the means of transportation concur. Transport buildings serve people and commerce, but do so where the train, bus, boat or plane decides. Stations require access to tracks, and tracks to trains, and trains need to stop and pick up passengers and to refuel. Ultimately, it is the train that decides the spatial geometry of the transport system, not the existence of people or urban areas. Trains are taken into cities and, where the station occurs, so too does growth and prosperity. Where trains are required to stop to pick up crew or to refuel, sometimes in the middle of nowhere, towns have grown up in the past, such as Swindon in England and Dortmund in Germany. Airports, too, exist where planes have to be refuelled and the air crew replaced, although in the past their location often had military origins. Around such airstrips, new cities also eventually grow, such as at Heathrow and Kastrup.

Just as road transportation grew on the back of mechanical skills and entrepreneurship generated by the First World War, so airports owed their initial conception and location to military needs. Wartime flying led to the development of civil aircraft in the 1920s and to commercial passenger planes in the 1930s. Since many early airports were shared between military and civil authorities (e.g. Croydon, Surrey, 1930), the arrangement was not without difficulty, especially when passport and customs controls were introduced. Eventually, congested airways, access to road and rail systems, and security concerns led to the separation of airports into separate civil and military establishments after the Second World War. However, the military culture survived in the lack of integration between transport modes and between land use functions.

The rise of inter-modality

The crossing of the paths of trains, buses and planes was once thought an irrelevance, but today the concept of inter-modality brings the transport systems together in the interests of society at large. Integration of transportation can be traced as a trend in the twentieth century, in spite of the cross-currents of competition and fragmentation that often occurred. Only recently have the advantages of an integrated transport system been acknowledged in official government thinking, particularly in the UK. To arrive at an airport by train, having previously caught a local bus to the station, provides benefits for the public and leads to fossil fuel reduction, though less tangible benefits accrue for the transport companies involved. Profit requires a degree of 'dwell time' at the station or the airport, which integration often denies. Over the past hundred years, competition has led to inefficiency and the redundancy of infrastructures. Though the Beeching Report made obsolete about 40 per cent of the UK's railway network, it did little to promote an integrated system that effectively interfaced with buses and planes. Recently, however, buildings for trans-portation have learnt to acknowledge the benefits of bringing all transport modes together. So, instead of the railway station,

bus station and airport, we have 'interchanges' – places where the lines and corridors of movement coincide. The formal consequence of the hybridisation is the emergence of a new type of transport building. The modern interchange (Moscow, Arnhem or Berlin central stations are good examples) is part station, part departure lounge, part bus concourse. The internal imagery draws upon different design influences, yet, in spite of the complexity, these are still recognisably transport buildings.

What many modern interchanges lack, unless they are adaptations of existing urban railway stations, is a sense of location. Their position is the result of an abstract play of lines drawn upon the face of a city or its surrounding countryside. But there is a placeless beauty in some of these megastructures – a sublime scale and sense of engineering, whose roots take us back to the heroic origins of nineteenth-century railway stations. The denial of place, of season and of geography allows the modern interchange to explore space, movement, light and construction as the primary elements of building character. No longer is style a metaphor for an externalised image – these new stations cum airports are the very essence of transport architecture, using a deep and enduring language of technology, dramatic movement spaces and architectural forms.

The authenticity of the shared concourse, rather than the superficiality of commercial space, of structure that honestly expresses the laws of physics, and well-lit volumes that sparkle in sunlight provide an image that takes transport architecture into the twenty-first century. One has only to walk through the new St Pancras/King's Cross Interchange in London to experience the rebirth of a type. Here, the extended station (called St Pancras Eurostar International) is the reforging of values acted out in architecture and engineering, whose origins lie in countless examples across Europe.

Transport architecture is torn between the utilitarian, the pressures of economic sustainability and the romantic. For many, the station or terminal is a functional building that provides the connection to train, bus or plane. For others, it is a place to welcome friends and lovers – the backcloth to more emotional moments. For others, too, it provides an alternative vision of a public realm where shared values and collective movement matter more than being cocooned in private vehicles. The volume of the building acts as a container for memories, as well as providing the functional means of access to cities or trains, and by implication also to sustainable development. This is perhaps why the building type has evolved into an interior of height, volume, daylight and natural ventilation.

2.12 Remodelling of Birmingham New Street Station to better cater for traffic flows. The design by Foreign Office Architects expresses modern interchange values. (Courtesy of FOA)

The station, and travel concourse in particular, is the arena of memories, of encounters and goodbyes, and also of future visions. As such, it is a spacious container whose dimensions go beyond that which can be justified by utility alone.

Elements such as bookshops, cafés and flower stalls reinforce these non-functional connotations. Even architectural design engages in a dialogue with the romantic or irrational – gothic cusps and classical swags bedeck the structure of old terminals, and high-tech assemblies new ones. The typical station is rich in imagery that is often a far cry from the functionality of the trains and buses. Even airport lounges, with their tree-like columns and umbrella roofs, hint at an uplifting moment, or memories of a lost nature.

In large airports, as many people are shopping or waiting for friends as are queuing for flights. The non-travelling public drawn to airports by shops and leisure facilities creates a collective audience that challenges the basis of the airport as solely a gateway to the plane. The complexity of people needs and the diversity of transport modes are reflected in multi-layered interactive environments. The human richness supports functional diversity, which in turn feeds the multi-use and coding of space architecturally. The initial simplicity of design gives way over time to plurality, as inter-modality takes over. Order beloved of system planners, architects and engineers erodes into rather messy complexity. It is the social interactions on the ground, combined with spectacular feats of engineering overhead, that are the essence of character of many transport buildings – old and new.

This character is most prevalent in the booking halls and public concourses of transport buildings. It is less evident in the platform and gate lounge areas. Here, smaller and more linear volumes prevail – long spaces designed for queuing or waiting for the line of carriages that make a train or the long fuselage of modern aircraft. The interaction between contained, yet spacious, volumes and the linearity of platforms forms the basic chemistry of the building type. Added to this, there are often bridges or tunnels to cross the railway tracks. The three elements – hall, platform and bridge – are repeated in the airport in the form of concourse, lounge and gate. It is the changing relationship between the three that helps identify transport buildings and that provides a clue to the evolutionary development of the interchange.

Early stations were merely sheds that protected passengers at the point where they purchased tickets. Platform canopies occurred later. As the type matured, the platform dominated the concourse, except at larger stations and terminals. The glass roofs and iron framing of the platform gave character to the station. In time, the station was often merely an enclosure of platforms that adjoined an understated ticket hall. Within the platform area two theatres are acted out – that of the train and that of the passenger. Often, but not always, the two would meet for a moment. Then the platforms are a landscape of flashing columns, name boards and bands of light and darkness. What makes the interchange exciting is the way this fundamental theatre is amplified into ever greater drama as the number of connecting transport services increase and hence the flow of people is multiplied.

So, whereas the booking hall and its surrounding flower stalls, bookshops, burger bars and cafés are the places for human gathering, the platform is the location of more mechanical dramas, where people come into contact with the means of movement. The same is true of bus stations, ferry terminals and aircraft terminals. Some designers, such as Foreign Office Architects in Yokohama, have exploited these ambiguous qualities. Normally, it is people who instil meaning into often neutral volumes – the space is an open frame converted into theatre not by architectural means (though there are exceptions), but by the passengers themselves in their complex movement and social interactions within the interchange. It is people, not commerce, who give meaning and ultimately justification for public transport buildings by their interaction with the world of speed, technology and architectural space.

However, the pure volume of people and their needs for snacks and reading material has led to the emergence of the shopping mall concept within concourses. The worst examples require passengers to dodge between retail outlets as they seek to gain tickets and travel information or board the train. Here, the train (or plane, bus or ferry) is at the far end of the shopping centre – barely visible within the chaos and advertising noise of the mall. Ideally, retail should be around the interchange, colonising the external spaces and directing people along the edges towards the transport network. Where retail is inside the interchange, the concourses should not be impeded, either

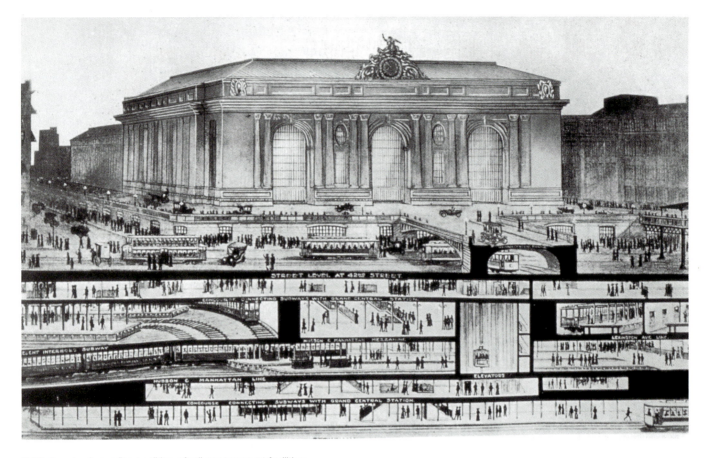

2.13 America has a fine tradition of railway transport facilities. Today they are often neglected and poorly connected to modern urban life. Grand Central Station, New York, in 1904. (From author's collection)

physically or in terms of route perception. Where shops are inside, they should not be in the way and neither should advertising hoardings, banners and queues obstruct sight lines.

One problem with older transport facilities is that earlier design guidelines are either lost or ignored. This results in management not having a clear idea of the original architectural purpose, such as the preservation of viewing lines and space corridors. Since transport buildings are very long-lived and subject to great pressure for change, the impacts of incremental change over time should be anticipated at the design stage and reinforced through management agreements. A good example of managing the dual interests of the passenger and retailer is Paddington Station in London (the new interchange for the Heathrow Express), which exploits the cross-section particularly well. Here, there are decks that contain shops and bars overlooking the main concourse, while around the edge of the concourse retailers such as Sainsbury have outlets that face into the station and out to surrounding streets. This is achieved within a grade 1 listed building that also connects to underground railway lines, high-speed and regional train services, and surface bus routes.

Innovations in the USA

The Americas are an unlikely place to find cutting-edge new transport technologies and associated interchange facilities, yet innovations in light rail and bus transportation are setting examples for other nations to follow. There are two reasons for this: first, the expanding communities of the south and west of the USA (which house many elderly people and immigrants) are putting investment not into roads but into bus and tram systems. This is driven by fears of rising petrol prices, and concern over congestion and air quality (Wortman, 2005: 2–8). Second, in the established cities of the north and east, under-investment in public transport has left an inherited backlog of decaying infrastructure that politicians are keen to address. Here, new transport technologies are being employed alongside well-established transport networks. In both cases, the car is not eliminated but is placed into a hierarchy that has integrated public transport at one end and personal car-based mobility at the other. As American cities decay and choke on the effects of unregulated ambition, there has begun a new recognition that successful civic cultures embrace public transport as a full partner within the transportation infrastructure. From Los Angeles to Boston, well-integrated, efficient and cost-effective public transport networks are seen as boosting local economies, improving quality of life (particularly for those in the inner cities) and raising civic pride (ibid.).

2.14 Health concerns over air quality and obesity form one of the drivers for change in US policy on transportation. Walking, cycling and trains are now important elements of the drive towards sustainable means of mobility. (Photo: Brian Edwards)

In this change of emphasis over the benefits of public transportation, American cities have also developed new approaches to interchanges. Since the car remains a dominant force, the kerb-to-platform relationship is important, and since movement by long-distance train is still relatively rare, the interface between bus, tram and heavy rail is not as developed as in Europe or the Pacific region. However, at the fine-grained community level where car, bicycle, bus and light rail interconnect, there are many innovations to be found. These concern integrated multi-modal links to new business districts, and the development of new communities and public transport as combined packages, both financially and in terms of urban design, and in providing special interchange facilities for the elderly. Since the latter are expected to outlive their driving capacity by several years, public transport providers have been quick to innovate on behalf of America's ageing population. There has emerged, therefore, a new consensus that economic prosperity, social welfare and environmental conservation are effectively addressed through the arena of public transportation. It was a theme early in the Obama administration, as America began to restructure its economy around green jobs and joined the international league of nations concerned with climate change.

In terms of the interchange, the use of hybrid diesel and electric engines has improved efficiency, reduced emissions, especially of toxic particles, and cut down on noise levels. These have allowed the close integration of transport systems with the home, workplace, leisure and education facilities. Fine-grained integration of communities with light rail networks is prevalent in many big and medium-sized cities. Here, the connection is by foot, bicycle, car and local buses, using shared civic space rather than dedicated roads or tracks. Hybrid vehicles lead to hybrid space use, which favours those walking and cycling while also subtly downgrading the dominance of the car. This, in turn, has encouraged the development of large paved plazas around transport stations, which encourages the sense of shared space and promotes an air of civic democracy. In this, the Jeffersonian ideals of equal access to society's riches are acted out in terms of urban space planning.

The elderly have benefited from the resurgence of investment in public transportation. Usage levels by this group have expanded faster than expected, leading to further investment, which in turn has led to new users being drawn out of their cars. One key to this is seen as the quality of station and interchange design, and the type of facilities provided, including web-based timetabling and integrated information at the bus stop or station. Real-time information rather than the abstractions of a timetable are important for user satisfaction of the

2.15a–b The interchange has to accommodate changing modes of personal transport: (a) electric car, (b) new family bike. (Photos: Brian Edwards)

interchange as a whole, particularly when email is used to alert passengers of delays (Wortman, 2005: 2–8).

Part of the quality profile is that of safety and security at transition points. American interchanges do not feature the arrays of security cameras found in the UK, but the attention to lighting, the overlooking of all public areas, and the generously dimensioned level or ramped approaches give modern transport facilities an air of comfort and safety. Added to this, storage for bicycles and their transport on buses, ferries and trains are considered essential if inter-modality is to work.

Public transport is also seen as helping to promote fitness and overcome obesity among American people. Hence, there is an emphasis on muscle power, in the form of either walking or cycling, at interchanges. Many of the funding programmes that have supported the expansion of light rail in new suburbs or on American university campuses have done so on the basis of integrated benefits across social, environmental and economic fields. Health costs associated with lack of exercise and poor urban air quality have been used to justify a switch

in funding from road to rail and from traditional polluting buses to intelligent hybrid ones.

While federal government and local civic leaders have begun to see the wider benefits of investment in public transport, the area of regulation has also been tightened. After years of deregulation that favoured the automobile industry, there has emerged recognition that public control and urban planning are key ingredients in the provision of successful low-carbon cities. The failure of deregulation elsewhere (for example in banking) has revitalised the role of the state in meeting community demands in key areas such as health, environmental protection and, critically, in transportation. Mobility has begun to emerge in US cities as a political issue and this has spurned the growth in light rail and revitalised traditional heavy rail and bus provision. Political activism and public transport are closely related in many American neighbourhoods, making the completion of transportation projects a source of considerable civic pride. Thus, the new interchanges that have emerged at the crossings of older networks are emblems of rebirth among the decay of the typical inner city.

PART 2

Design strategies

Having explored some of the concepts and histories of mobility and the interchange, Part 2 sets out the design implications at an urban, building and interior level. The aim is to guide the architect towards better serving the needs of the public, while also meeting the demands of clients. A key here is the integration of professional skills – urban planning, landscaping, architecture, and civil and structural engineering. Transport architecture is shaped by the strict demands of infrastructure engineering, of cost and of many technical factors from safety to energy efficiency. It is also moulded by human and aesthetic values: the legitimate demands for spatial clarity, ease of orientation, sense of uplift, speed of movement in and out, and design that reduces stress. How the technical, human and mobility needs are reconciled in the arena of transport architecture is the central theme of the two chapters in Part 2.

Chapter 3 explores design strategies mainly at the urban design level. It deals with integration between mobility and city form, discusses different geometric configurations common to successful transport hubs, and raises the now familiar debate between energy use and urban density (and hence the importance of transport interchanges). This chapter also discusses the role of the architect vis-à-vis that of the engineer and how their relative positions are changing under the impact of more people-friendly forms of urban movement.

Chapter 4 is more technical in nature and outlines particular solutions to common problems found in transport architecture. These deal with structural and environmental issues that occur in interchanges, as well as human factors such as disabled provision and the role of public art. The chapter also outlines strategies for the selection of construction materials, the design of lighting and signage, and how space is organised in hierarchical terms in both plan and section. The aim is to offer the reader a rationale for choosing particular design approaches to building and interior design.

Strategies for urban design

The interchange has evolved from earlier transport building types in response mainly to societal and environmental pressures. As a general rule, the modern interchange acts in the public rather than the commercial interest, since the integration of facilities and transport modes undermines the concept of dwell time (and hence retail profit) and a captive travelling audience that underpins more separated transportation systems. Also, as a general rule, the interchange is centrally located in urban areas where the wider economic benefits can be turned to social advantage for often less privileged communities. In this, the modern interchange is simply following the example of the stations of the first railway age, which created new urban quarters around the historic cores of cities. But as the social and economic benefits are felt, the city as a whole gains by reducing traffic congestion, air pollution and car dependence, and this in turn encourages the middle classes to return from the suburbs. Plans, therefore, for urban regeneration should not ignore the power of the interchange to bring about a wide range of environmental, social and economic advantages.

Urbanisation is good for sustainable development (Ritchie and Thomas, 2009). It allows goods, services, infrastructure and resources to be shared by the largest number of people. The more compact the city, the more sustainable it is in energy terms. It is no wonder that there has been a drift to cities over the past 50 years. In fact, it is predicted that, by 2020, three-quarters of the world population will live in urban areas. This is good for public transport and bad in the long term for those dependent upon the private car. However, urbanisation on its own is not sufficient – there needs to be investment in new ways of living, new ways of moving about the city, new ways of thinking about work, leisure and education, and new concepts of urban density.

One way this can be helped to happen is by seeing transport routes through the city as migration corridors. Here, mobility in social and economic terms can be promoted with assistance in the form of government grants for commerce or technological innovation being focused on these corridors. Here too, low-interest loans could be introduced in order to induce low-income families to live in the new public transport corridors of the future. The idea of 'migration' is about movement – economically, socially and culturally – in order to overcome inner-city poverty, using the interchange as an agent of change (Steffin, 2008). Since people in the USA spend roughly 20 per cent of their disposable income on transport (in Europe approximately 12 per cent), the preferences

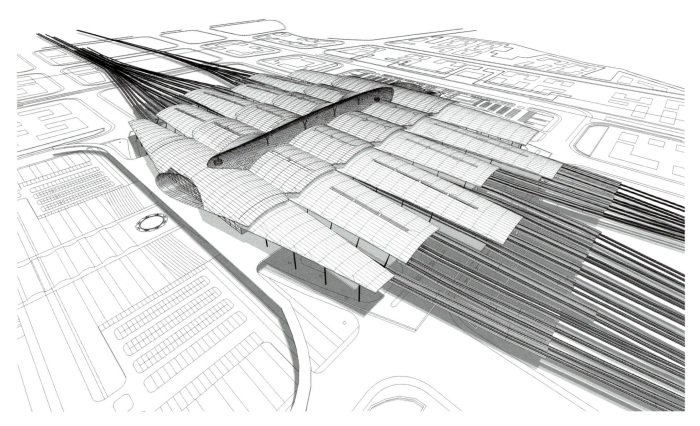

3.1 The massive Guangzhou Station design by The Farrell Partnership shows the role of the interchange concourse in bridging sections of the city divided by railway infrastructure. (Courtesy of TFP)

in income and loan support should be towards those areas well served by public transport.

Ensuring that urban areas are attractive to all classes and types of people is one of the prime targets of sustainable urban living. Making cities as attractive as suburbs as places to live and work requires investment in transport infrastructure, environmental quality and social provision. Better transport provision goes alongside better schools, colleges, healthcare, libraries and recreation facilities. An upgraded environment, less air pollution and congestion, and reduced crime levels are generally consequences of investment in public transportation. Government investment can change the marketplace of inner-city economics, leading to physical transformation and a better quality of life for all. After all, the concept of the compact city suggests a liveable urbanism built around public transport and, since public transport drives an increase in urban densities, it is one of the keys to achieving lasting sustainable development (Rogers, 2009). As densities increase, so too does the shift from singular provision such as railway and bus stations to integrated transport hubs. There is also a move back to mobility on foot, and by cycle, metro and bus. New modes of mobility will replace the car as the primary means of transportation. In turn, mobility shifts from

singular long journeys to multiple short ones, where interconnection between modes is critical in terms of user satisfaction. With the concept of interchange comes also the creation of new public spaces – areas of the city where people can meet, exchange ideas and transfer between transport modes.

Walking provides one of the main means by which transport interconnection is achieved, yet walking is often overlooked by transport providers and their designers. In the UK, 80 per cent of journeys under one mile are made on foot, although only 3 per cent of total travel distance is achieved by walking. Journeys on foot are the dominant travel mode for short journeys and these often entail movement between transport systems (Tolley, 2003: 72–3). Walking is not, however, undertaken at an even speed. Different people have different levels of mobility and fitness – hence at interchanges the emphasis has to be upon level surfaces, smooth pavements and seating areas for the less mobile. Encouraging journeys to be undertaken on foot means fewer car and taxi journeys, and hence less pollution and also a healthier population. Unfortunately, walking is the least considered of the journey modes by those involved in infrastructure planning, yet feet are what binds together the typical interchange.

Interchanges exist at metropolitan, regional and local levels. Major city centre interchanges bring about different environmental, social and economic benefits from those of

3.2 External plaza at Birmingham New Street Station, as designed by Foreign Office Architects. Notice the priority given to pedestrians. (Courtesy of FOA)

suburban locations, and also raise different design issues. Also, different types of transport interchange lead to different potentials in terms of sustainable development. For example, social sustainability is more likely to be achieved in inner-city locations or where transport interchanges are located near to large housing estates. By way of contrast, city centre train-based interchanges will enable new economic potentials to be realised, while suburban bus interchanges with park and ride facilities have their justification in the arena of reducing fossil fuel use and air pollution. It is important, therefore, when designing interchanges, to consider specific benefits to the wider community rather than apply a universal solution. This relates not just to the location of transport facilities, but to the preferred mode of travel to the interchange and between its parts. Hence, in some locations buses and taxis glue together the web of interconnecting transport services; in others it is the bike and journeys on foot.

The bicycle is commonly seen as one of the most important transport types for the post-fossil fuel age. In many developed countries and in emerging economies the bicycle is both a quick form of mobility and a cheap means of transport, and provides that combination of health and well-being characteristic of the twenty-first century. Cycles now provide cross-city movement on an unprecedented scale, and constitute one of the most important interconnecting transport modes at urban interchanges. Part of the growth in cycle use stems from new forms of cycle ownership. Cycle clubs are growing up to mirror urban car clubs. Rather than people owning a bicycle, these clubs provide sufficient bicycles to allow the cyclist to find one on the street and take it over for the duration of the journey. The bike is then left for the next user. In Paris, a scheme of this nature has attracted over 200,000 cyclists, who pay a nominal annual club membership fee. The user does not have to lock the bicycle or find a parking area or bike store for it during the day,

3.3 Cycle storage at Aarhus Station. (Photo: Brian Edwards)

which is one of the key attractions. Such initiatives increase cycle use at the expense of car, especially where car pricing acts as a further incentive, as in London and Paris. New patterns of cycle use coupled with investment in new transport infra-structure promise to reshape the geography of many cities both in Europe and in Asia.

In many ways, investment in modern public transport is investment in smart infrastructure, as against inefficient and more brutal investment in roads and provision for cars and lorries. The compact, socially diverse, economically active and culturally rich city of the future is one that sustains alternative forms of transport. From movement on foot to cycle, to bus, to metro and then to train requires integration of facilities. Mobility in the city of the twenty-first century is a question of integrating a diversity of transport modes, ensuring that the interfaces work well. Basically, the big spatial intersections of the city are where the transport intersections occur and this is where the interchange resides.

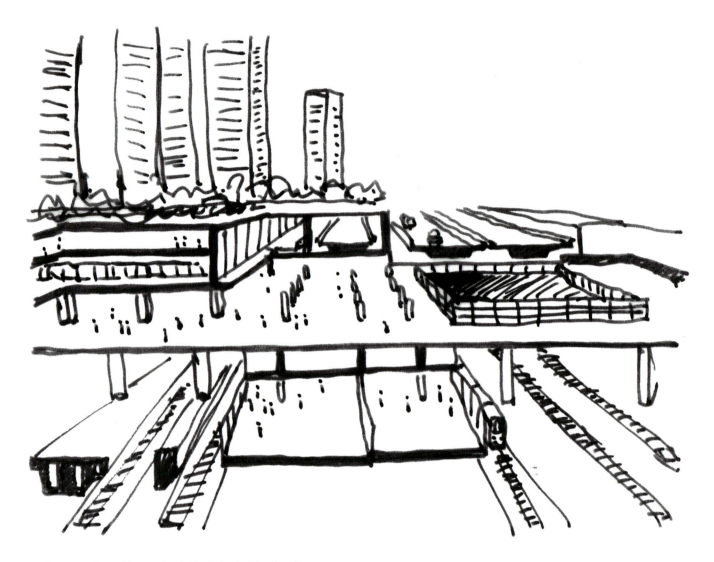

3.4 Sketch for Nam Cheong Station in China by The Farrell Partnership, showing the multi-levels required in typical interchanges. (Courtesy of TFP)

Besides existing modes of transport, the rising urban problems of the twenty-first century will probably generate new forms of public transport, or radical developments of existing forms. This will include innovations in light rail, taking the kind of technologies seen at airports and theme parks into towns and cities. Light rail is relatively inexpensive and can thread its way along or above existing streets with ease. Changes in level and incline can be accommodated with much less cost than heavy rail, and numbers of passengers carried can greatly exceed those on bus systems. Added to this, there are levels of comfort and fuel efficiency with light rail that few other modes can match. There are also likely to be developments in monorail systems placed at high level above existing street space, where transport systems are vertically stacked (Vancouver's Skytrain, though based on light rail technology, shows the potential of elevated monorails). Another innovation is that of moving pavements to ease pedestrian mobility in city streets.

As these technologies develop, the distinction between the inside and outside of buildings becomes blurred, since monorails and moving pavements will need to weave their way through shopping malls in order to attract sufficient numbers to justify the initial outlay. As with all transport modes, it is vital that transport planning and urban planning are integrated, and that development corridors of sufficient density are created to support the transport infrastructure costs.

Transport modes are moving from familiar to less familiar forms. Buses now come in extended formats capable of carrying over 200 people. Buses also use guide systems for extra efficiency and employ lowered boarding floors for disabled access. The latter saves the cost of special raised decks at bus stops. Trains, too, come in a wide variety of formats, from super high speed to local single carriage systems that operate rather like rural buses. Hybrid technologies are allowing for a number of innovations in transport design, many based upon mixing the genes of different transport types and transferring technologies from amusement parks to cities. Although

3.5 High-speed interchanges impact upon cities and regions.
Naples Vesuvio Interchange, designed by Grimshaw.
(Courtesy of Grimshaw)

technological innovation is important, so too is the quality of design of the transport modes of the future, if public transport is to attract users away from private cars.

Another innovation is in fuel cell technology, hybrid power systems and bio-fuels. These promise to reduce the carbon footprint of transport, improve air quality and reduce traffic noise. Since transportation systems are concentrated in urban areas, it is here that noise and air pollution is also focused. Hence, innovation in new fuels, in new traction systems and in new vehicle design promises to reshape transport and improve the quality of life for those in the major cities of the world.

In designing the interchange there are three geometries that require reconciliation – that of the city, that of the spatial logic of the transport systems incorporated, and that of people movement within and around the system. Rarely are these effectively balanced with all too familiar consequences, such

as the dislocation of internal and external routes or the reversing of obvious hierarchies in transport provision (such as the priority given to taxis over bicycles or pedestrian flows). However, as the number of transport modes increases, so too does the difficulty of planning the interchange as a coherent whole. When two transportation types meet, the planning of the interchange is relatively simple, but with three or four interconnecting modes the designer inevitably has to plan the facility in three dimensions. As space planning becomes more complex, so too does the management of time and hence timetabling of transport facilities. Organising the building in three dimensions and reconciling this with a city plan that is two-dimensional is by no means straightforward. Equally difficult is the task of orientating passengers who are involved in three-dimensional movement flows – to and from the transportation system and interacting between the various multi-level transport modes as interchange passengers.

Normally, users of transport facilities prefer single-level stations and interchanges. However, this is rarely possible with

Table 3.1 Characteristics of types of travel mode in European cities

Mode	Speed	Typical journey distance	Journey time	Modal utilisation (%)	Transport means utilisation (%)
Walking	3 mph	0.3–1 miles	5–20 mins	100	25
Cycling	6 mph	1–3 miles	10–30 mins	100	15–30
Bus	10 mph	1–5 miles	6–30 mins	40	25
Metro	15 mph	2–6 miles	7–20 mins	30	20
Train	20 mph	3–10 miles	7–30 mins	30	25
Water bus/ferry	5 mph	1–2 miles	12–25 mins	15	2–4

Table 3.2 Characteristics of types of travel between European cities

Mode	Speed	Typical journey distance	Journey time	Modal utilisation (%)	Transport means utilisation (%)
Coach	45 mph	45–90 miles	1–2 hours	60	10
Regional train	50 mph	25–100 miles	30 mins–2 hours	50	40
High-speed train	120 mph	100–250 miles	50 mins–2 hours	50	45
Plane	300 mph	300–600 miles	1–2 hours	60	40

large multi-modal interchanges. Passengers also want clear and legible spaces supported by clear travel information. Light is also important both to the legibility of facilities and to the pleasure of using interchanges. Underground stations and interchanges pose a particular problem in directing natural light into subterranean locations. Light wells, light walls and reflective materials can be employed, but the deeper the platforms and concourses, the harder it is to utilise natural light and ventilation. There are consequences here for energy efficiency.

Criteria for the design of transport interchanges are:

- An interchange should be an uplifting experience.
- There should be interest at the arrival and departure points.
- There should be a welcoming environment.
- Distinction should be removed between front and back.
- Natural light should be maximised and physical enclosure reduced.
- Good air quality is essential above and below ground.
- Sight lines should be maximised to aid passenger navigation.
- Transport modes should be seen, not imagined.

(Adapted from UK Highways Agency,
2002, cited in Blow, 2005: 16–17)

Integrating the interchange with the city

Transplanting new interchanges into existing urban areas is by no means easy. Transport buildings have always created barriers to movement. Railway tracks, bus routes, airport runways, tunnels, bridges and tramlines impede the free flow of pedestrian and cycle traffic as well as private cars. The permeable city becomes impermeable in the vicinity of major transport facilities. Restricting the barrier effect of infrastructure

is the task of the city planner rather than the architect. However, the design of the building can do much to aid cross-city movement, by opening up the interchange to all users rather than just the travelling public. This suggests the creation of a major space, enclosed or open, through which people can flow or gather as they wish. The emphasis here is upon the public rather than the transport system and upon movement by foot as against that by wheel.

The idea of cross-city movement is enhanced if concepts of front and back are removed. The interchange is a structure of multitudinous movements, of threads of people and infrastructure that are joined loosely rather than tightly. This suggests space and light and movement. Movement in all directions opens the city to potential economic growth corridors – areas that were once seen as 'backs' become desirable fronts with the interchange, as at Melbourne South Station, generating urban reorientation. This can benefit communities once cut off by railway tracks or blighted by low-grade industry.

Putting social objectives before commercial ones also helps to restrict barriers to movement. Unfortunately, the usual pecking order is that of the transport infrastructure and all of its engineering structures, followed by commercial and retail interests, with people left to navigate as best they can through disjointed facilities. Placing the emphasis upon space rather than infrastructure helps to rebalance design decision making. Such space can be either an external square linked, on the one hand, to the pattern of city streets and, on the other, to the various transport systems or modes available at the interchange. Alternatively, the space can be enclosed beneath a roof, making a huge market hall full of people and transport services. The latter solution is increasingly the model adopted, making modern interchanges and their counterparts at airports among the largest enclosed spaces in the world.

3.6 Transportation needs to address safety issues at an urban design level. Road crossings near trams are particularly hazardous. (Photo: Brian Edwards)

Whether the interchange is an outside urban space or an enclosed inner one, the task remains of joining it with existing movement patterns in the city. It helps if interchanges are positioned along the major road system, where the hierarchy of streets and that of public transport can complement each other. Such an arrangement gives prominence to the interchange and provides easy access by car, by bicycle and on foot. However, road space will have to be taken from car users to accommodate other modes of surface transport. The creation of cycle ways, wide promenades for pedestrian movement, and possibly tramways will stress the existing road system around the interchange. Visual prominence and a well-functioning road system for all members of society should be the aim. Transferring some road space to pedestrian-only routes also helps by increasing safety for pedestrians and by removing barriers and obstacles such as roadside kerbs. Since many users are likely to be old or parents with children, surfaces beneath feet should be smooth and have even gradients.

Key ingredients for the successful integration of interchange and urban areas are:

- safe and secure routes for pedestrians;
- safe, secure and prioritised cycle ways;
- integration of commercial space at interchange with hinterland;
- formation of civic space around interchange;
- creating interchange spaces for use of wider community;
- ensuring legible connections with other transport modes and civic landmarks.

Defining the presence of the interchange through urban design is both a visual and a practical skill. It should be possible to view the interchange from the major streets and, here, it helps if a square is created at the perimeter of the interchange to give further definition to the transport hub. Such a square operates on two levels: it gives civic status to the interchange and it provides a gathering space for users of the various transport facilities. The square (it could be formal in shape or more organic depending on the nature of the urban area served) also functions as a space to guide passengers and to aid navigation. Here, there are likely to be interconnecting surface transport systems such as bus and tram, and routes beneath ground to

3.7 Sketch for Mei Foo Station in China by The Farrell Partnership, which integrates transport and landscape within Lai Chi Kok Park. (Courtesy of TFP)

3.8 Clarity of urban space and smooth finishes underfoot are features of the forecourt to Oslo Station. Notice also how well old and new are integrated. (Photo: Brian Edwards)

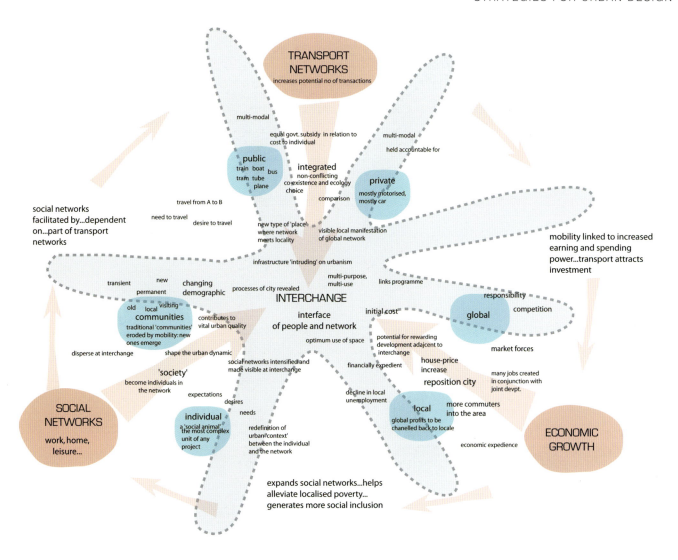

3.9 Fiona Scott's model of the qualities needed to make an effective transport interchange. (Courtesy of Fiona Scott)

Table 3.3 Problems and design solutions typical of the UK's transport interchanges

Problem	Design solution
Disconnection	Urban connectors
Accessibility	Access ribbons
Poor public space	Community hubs
Low density	Stacked programmes
In-between spaces	Residual space-markers
Poor information	Info-flow

underground railways. Stitching this space into the network of existing streets requires attention to movement hierarchies as well as the spatial pattern of city blocks and land users. Since the interchange will have a major impact on the immediate economic hinterland, there is likely to be further redevelopment over time. This suggests that city plans need to be updated periodically to achieve further refinement of the pattern of streets and blocks around the interchange. Here, the wider urban plan should pay particular attention to movement permeability, visual corridors, land use compatibility and future growth around the interchange. As Sir Terry Farrell notes, 'changing modes, management and the availability of transportation are reshaping our sense of urban space' (Wong, 2008: 5). This reshaping is an incremental process and flexibility in land use planning is required to ensure it enriches city life.

In her doctoral thesis at the Royal College of Art, Fiona Scott made a special study of transport interchanges in the UK and

Stratford Station in London in particular. Her work, published under the title *InterchangeABLE* (Scott, 2003), identified six problems and related design solutions associated with Stratford Station and typical of the UK's transport interchanges. These are set out in Table 3.3.

The emphasis in Scott's work is upon the pedestrian experience and on wider questions of social well-being. Her ideas are of value, particularly in the arena of integrating

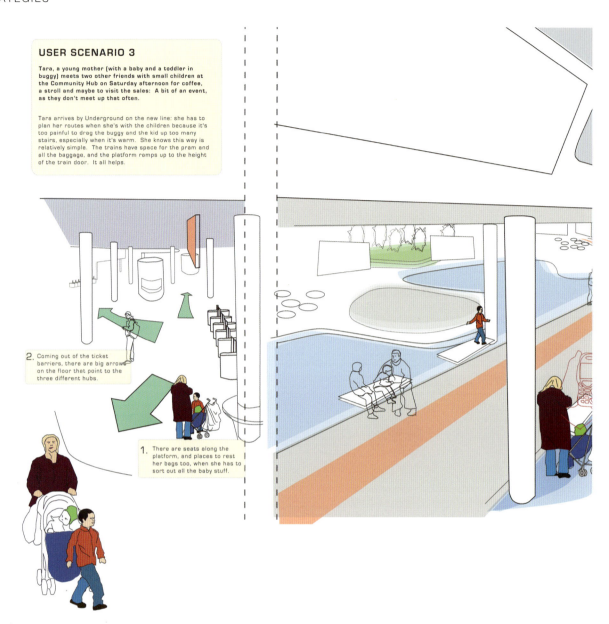

USER SCENARIO 3

Tara, a young mother (with a baby and a toddler in buggy) meets two other friends with small children at the Community Hub on Saturday afternoon for coffee, a stroll and maybe to visit the sales: A bit of an event, as they don't meet up that often.

Tara arrives by Underground on the new line: she has to plan her routes when she's with the children because it's too painful to drag the buggy and the kid up too many stairs, especially when it's warm. She knows this way is relatively simple. The trains have space for the pram and all the baggage, and the platform ramps up to the height of the train door. It all helps.

2. Coming out of the ticket barriers, there are big arrows on the floor that point to the three different hubs.

1. There are seats along the platform, and places to rest her bags too, when she has to sort out all the baby stuff.

3.10 Fiona Scott's drawing of the qualities needed of public spaces at interchanges. (Courtesy of Fiona Scott)

transport investment with that of urban regeneration. They span from city planning to the design of information boards, and from questions of economic benefit to those of social cohesion.

One problem of integrating linear transport facilities with the city concerns aesthetics. The tendency until recently was to standardise the stations along a line in order to provide consistency and reduce costs by removing variation. Hence, stations above ground and metro stops below were often designed using modular components and, therefore, a repeating language of materials, finishes and structural systems. Although this helped with branding and the identification of the transport network, it denied any expression of the different qualities that different neighbourhoods had. Today, the practice

is one of standardising some elements while giving architects the chance to adopt variety in specific locations.

This blend of modularity and particularity allows stations at different locations to express something specific about the neighbourhood served. It also allows an interchange to employ a visual language that is different from that of other stations. By separately articulating the transport hierarchies involved or by allowing the cultural nuances of locations to influence specific station designs, the system as a whole develops into a richer experience for the traveller. Employing cultural references helps with the integration of the abstractions of transport infrastructure with the social complexities of urban areas. This may allow a station or other form of transport facility

to become 'owned' by the community, thereby reducing vandalism or graffiti. It is an approach adopted on London's Jubilee line metro stations and the light rail stations built for Hanover's Expo 2000.

Arup's model for the interchange

In 2008 Arup, one of the world's major engineering practices, compiled a set of operating themes to guide its own thoughts with regard to the planning and design of interchanges (Dobrovolsky, 2009). The themes are:

- accessibility;
- operation;
- constructability;
- sustainability;
- liveability;
- phasing.

They form a template of issues that are discussed by Arup designers at the start of projects with clients, the participating architects, planning authorities and users. Each is worth examining.

Accessibility: This deals with physical access in the form of feeder routes (walking, bike, bus etc.) to the interchange, but also questions of land ownership and legal obstacles to creating smooth movement patterns are raised.

Operation: Here, Arup explores inter-modality, funding partnerships and the role of retail at the interchange. Ensuring that the economic basis is sound and, particularly, that a good public/private financial interface occurs here.

Constructability: This has a bearing on cost and the timetable of operations. Over-complex solutions are identified at the outset and remedies put in place to ensure the project is buildable within budget constraints.

Sustainability: Energy and environmental impact are growing areas of public concern and regulation. They are addressed at the outset and are broadly framed to ensure that social, economic and environmental sustainability is bedded into projects at inception. Arup sees this as a key area for the interchange to address.

Liveability: Questions of design quality and social provision form the core of interests here. The concerns addressed by

3.11 Fiona Scott's drawing of the qualities needed of access routes leading to transport interchanges. (Courtesy of Fiona Scott)

3.12 Interchanges impact upon a range of urban, environmental, economic and social issues. They are one of the most complex design problems and require close cooperation between architect, planner and engineer. Lisbon Oriente Station. (Photo: courtesy of Torben Dahl)

Arup under this heading often extend into the hinterland of interchanges to ensure that transport, civic design and feeder routes are considered. This often involves engagement with public agencies and potential developers of peripheral sites.

Phasing: Putting the financial package together to ensure that the project can be achieved over time is critical, since interchanges involve a number of infrastructure companies, public agencies and a diversity of sources of finance. Phasing involves putting together a stakeholder framework of financial commitments before the project begins.

Although the template is flexible, the framework it provides ensures that different cultural traditions in the provision of transport projects are accommodated within a robust armature. It allows, for instance, for Chinese projects, with their tendency to put infrastructure first, to be rebalanced using wider sources

of funding to pay for community provision (such as dentists and bike repair shops) at stations. Similarly, the themes when applied to projects in the USA raise questions that state planners do not usually have to address, such as the relationship between interior and exterior urban space. The themes also have the effect of ensuring that major infrastructure transport projects integrate with the wider city beyond the aim of enhancing mobility. They raise questions of economic potential, social provision and environmental responsibility. The latter includes the growing inclusion of renewable energy technologies at interchanges.

The geometry of the interchange

The pattern of the interchange depends upon the number of transport modes accommodated, their angle and whether they are positioned above or below ground. Generally speaking, transport systems that interconnect at right angles are the most

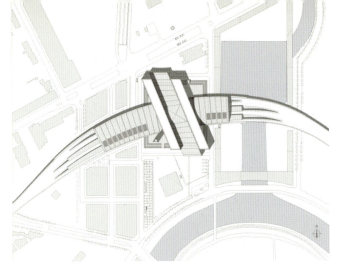

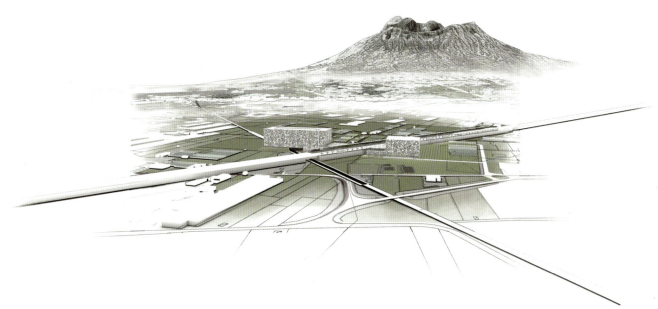

3.13a Geometric and urban considerations at the new Berlin Central Station, constructed on the site of the former Berlin Wall. Designed by von Gerkan, Marg. (Courtesy of von Gerkan, Marg and Partner)

3.13b The crossing of lines in the landscape formed by transportation is the starting point for this design by Grimshaw for the Vesuvio Station Interchange in Naples. (Courtesy of Grimshaw)

3.13c Model of Copenhagen's elevated metro line crossing the city, by students at Department 7, Royal Danish Academy of Fine Arts, School of Architecture. (Photo: Brian Edwards)

satisfactory for both users and designers. The engineering of right-angled connections offers both elegance and efficiency of structure. It is also readily comprehended by the travelling public. Parallel connections also provide legibility and design elegance. However, the difficulty arises when there are three or more transport systems interconnecting. Now three-dimensional planning is required as, for instance, when intercity train lines connect with regional and local lines, with buses and trams and perhaps also with underground metro systems. The geometry of surface and underground infrastructure, dictated often by existing configurations, has now to be reconciled with people movement above ground, people who themselves are following patterns shaped by the non-orthogonal geometry of the city. The key is to try to simplify the various movement angles so that the crossing of infrastructure paths follows the discipline of orthogonal geometry. This results in crossings in plan at 90 degrees and 45 degrees, and sectional arrangements that open up vistas of the different transport facilities available. Such an arrangement provides economy of construction, allows daylight and views to be exploited, and encourages comprehension of the totality of the interchange.

Necessarily, geometry, space and volume need to be planned carefully. This involves discussion at the earliest point in the project between the various engineers and designers of the different transport companies represented. More signifi-cantly, it needs a strong vision at the outset – one that is sufficiently inspiring to bring everybody on board. Here, geometric order is the key, not just in plan but also in section. However, it should not just be an abstract order dictated by transport infrastructure, but one that responds to the city both spatially and culturally. Also, since the medium of connection at the interchange is space, there needs to be sufficient investment in space, both qualitatively and quantitatively.

There is a pecking order in the formal functioning of an interchange. Normally, heavy rail systems take priority over light rail, rail generally has priority over bus provision, cars dominate cycle provision, and pedestrian needs tend to come last. Put another way, metal wheels win over rubber wheels, wide wheels over narrow wheels, and wheels over feet. The geometric arrangement of a typical interchange reflects these orders. They can be justified on the basis of infrastructure cost and the inflexibility of rail over bus transport. But the

consequence is that people tend to come last in design decision making, in spite of the obvious reality that the transport system serves people not infrastructure. A similar situation exists at airport interchanges, where the alignment of runways and taxiing areas for planes dictates the configuration of terminals. Even inside the terminals, the routes to connecting stations and bus services tend to be obscured by baggage handling systems on arrival and by security on departure. It is perhaps only the architect who can invest in the interests of the general public in an environment where so many competing infrastructure priorities exist.

Key factors in the geometric ordering of the interchange are:

- Seek to place major transport modes at right angles to each other.
- Place upgraded transport systems alongside existing ones.
- Ensure visual order follows functional order.
- Give priority to passenger geometry.
- Create ordered and legible spaces within the interchange.
- Use external landmarks as guides.

Geometric order needs to mirror passenger movements rather than just the transport infrastructure. Normally, an orderly configuration of interconnecting transport systems leads naturally to a movement sequence for passengers that is readily comprehended, efficient and pleasurable to use. The dictum that 'form follows function' should be modified at interchanges into 'form and function follow movement'. The prime function of an interchange is to facilitate connection to the transport infrastructure – hence, function is largely a question of connection achieved by people movement. Geometry and associated spatial hierarchies should prioritise space for movement, not space for lesser functions such as retail areas. Such movement is on foot for the most part and is achieved directionally. So the designer has two main tasks: first, that of providing sufficient space for the volume of movements entailed; and, second, that of ensuring that the sense of direction is maintained. In this, geometry alone is insufficient – there need to be clues to the direction to take. Such clues exist in maintaining visibility of the desired transport mode, in making it obvious where the city is to be found by maintaining views of landmarks, and in using daylight and sunlight to orientate passengers. Signs, though important, are less valuable than a well-ordered environment for the passenger.

In many ways it is advantageous to plot the desirable movement paths for people before the lines of transport infrastructure are established. This may involve wider questions of cross-town desire lines as well as plotting movement corridors for travelling passengers. Once the needs of people are established, the different transport modes can be traced on the ground. Giving priority to the function of people mobility alters the space/form relationship both inside the interchange and around its perimeter. However, elements of the transport infrastructure may already exist, or their future paths are fixed by engineering considerations. Then a compromise will have to be made between people on foot and wheeled systems. In this compromise it is imperative that the social dimension is safeguarded in a world where engineering and economic considerations are paramount.

People and their needs

The millions who pass through large interchanges are not a homogeneous group. There will be young and old people, those with small children, those in wheelchairs, those with limited vision, those seeking shelter (rather than travel), those using

3.14 Large transport interchanges, such as at the Channel Tunnel in Kent, impose their own geometry on the landscape. The flowing lines can lead to new design approaches. (Courtesy of BDP)

interchange shops for local produce, those as commuters who know the routes and facilities well, and those who are tourists and visitors who do not. The routes through the interchange that they take are likely to change level, change direction, and go from light to dark and from noisy to quiet. Generally, more poor people travel by public transport than by private means, more old people than young use public transport, and disabled people use more buses and trains than private cars. The role of public transport in supporting social inclusion, in maintaining quality of life for those disadvantaged, and in providing affordable travel facilities, should not be underestimated. Also, as more of the population becomes elderly and as more people move to big cities from rural areas, the public transport network assumes ever greater social importance. It is against this background that the modern interchange functions in a changing world.

The geometry of people movement is fundamentally different from that of the transport systems within the interchange. Although planes, ferries, trains, buses and trams have their own timetables, which to a large degree influence passenger flows, people are also independent of schedules and travel more randomly through available spaces. There are, therefore, two geometries of people movement – that of passengers intent upon boarding and that of the informal traveller. The former move in fast columns, the latter more slowly and multi-directionally. Space with its own design characteristics has to be provided for both. The situation is further complicated by the cross-currents of movement that occur. Various geometries collide as the paths of people columns cross. This is inconvenient for the able passenger, but alarming for those who are less able-bodied. Spacing movement in time helps overcome conflicts, but both time and space need management. The architect can only manipulate space and, here, the challenge is that of achieving legible routes and comfortable areas in which to dwell between taking journeys.

Passengers come in various shapes, sizes and types. They have varying degrees of mobility and the trend in the west towards obesity means that passengers are larger than ever. However, what is often overlooked are the different profiles of passengers and hence their differing perceptions and needs. For example, a commuter will know the sequence of interchange spaces well and will expect to pass through them quickly. On the other hand, the tourist will need to search for signs and may have to ask for guidance. The business traveller may expect a higher standard of provision and perhaps more privacy. Parents with children also have different needs, as do the elderly. Designing only for the rapidly flowing commuter reduces the standard for all. Hence, architects should insist upon quality so that minority as well as majority needs are met. Providing high space standards, seating and finishes not only serves the tourist, elderly or business user, but also enriches the experience of the commuter.

What makes an interchange different?

One of the big differences between modern interchanges and the earlier generation of railway and bus stations, and also ferry terminals, is the extent of gathering and movement space. This occurs internally and externally. In the interchange, the proportion of movement and dwell space is greater than in singular transport provision. The need to interconnect between transport systems, with inevitable time delays and enforced waiting, results in the need for extra space for gaining information, for simply dwelling and for crossing movement flows. Interconnection for multiple movements is more demanding of space than single movements. Hence, the geometries of space and the size of spaces become larger and more complex. Although the boarding areas such as platforms remain relatively constant, it is the movement zones that are put under stress and have to increase in dimension. Also, the bridge and tunnel areas, which have their own identity in single-transport buildings such as stations, become merged at interchanges. Now they are part of larger volumes, a set of sub-spaces within big public concourses.

One result of this is the extent of spatial and social interactivity within the major movement zones of the interchange. All spaces are connected to a degree in the pursuit of efficient interchange and are often celebrated architecturally. These large common movement areas embrace and integrate the sub-territories required to serve each individual transport type. Such big spaces become the main means of identifying the interchange as distinct from a bus or railway station. They may take the form of circular spaces as at Shanghai South

Station Interchange, or square ones as at Stuttgart Central Interchange. Usually, too, with interchanges of this size, there are linked interior volumes and exterior civic spaces. The latter provide space for pedestrian movements into the city, thereby forging connection to the economic, cultural and social realms. Interchanges have, therefore, big spaces that need to be celebrated architecturally, not merely provided functionally.

In many recent examples, the generosity and scale of urban transport interchange concourses begin to approach those of airports. They have much in common architecturally. However, two distinctions need to be made. First, the airport terminal does not form part of the city and, hence, urban connection via streets and footpaths does not occur. This results from the airport being largely a self-contained entity with all the problems associated with 'edge cities'. Few airports function effectively as urban interchanges and this affects the nature and function of the interior spaces provided. Second, the airport is rarely a true interchange, since the balance of facilities and their geographical isolation puts the emphasis on the airport function rather than that of interchange. Hence, the big democratic spaces associated with contemporary urban interchanges are rarely found at airports, where the concourses are linear and process passengers towards planes with only reluctant gestures towards rail or bus links. Multi-directional movements are characteristic of urban interchanges, not airport interchanges.

Stitching new transport systems into existing urban areas

The drive to reduce the 25 per cent of carbon emissions associated with transport around the world has led to the resurgence of interest in the benefits of public transport. Hence, governments and municipalities globally are investing in new rail and metro systems, many of which are threaded under or over existing urban areas.

Added to this, the impetus to support social sustainability and improve efficiency has led to the concept of the interchange – the place where transport systems are interconnected. Often it is the underground metro (Copenhagen) or street-level tram (Edinburgh) that provides the means of transfer between transport modes. In mature urban areas, interconnection involves considerable disruption and restructuring of the city. Although this is inevitable, two principles can be followed.

The first involves keeping the scale of the connecting vehicles to a minimum. Ideally, transfer is on foot, but this is not always possible. Often, bus, tram or heavy rail systems are needed to allow for inter-modal linkage. However, if the connecting vehicles are small they can be insinuated more easily into the existing urban infrastructure. Hence, it is better to use short trains that run more frequently than long trains that run less frequently. For example, the planned circle line metro

3.15 When new transport facilities are added to existing stations, it may be possible to convert older structures to parks and gardens, as here at Atocha Interchange, Madrid. (Photo: Brian Edwards)

3.16 Traditionally, interchanges had large gathering spaces outside their main entrances, as here at Amsterdam Central Station. (Photo: Brian Edwards)

system in Copenhagen is designed to run with three driverless carriages initially (later increased to four), running at one-minute intervals (this achieves a peak capacity of about 15,000 passengers per hour). Short trains mean short platforms, which entails less restructuring of the city above. It allows, for instance, for the existing street space to be utilised rather than having to tunnel under buildings. The same is true of tram systems, which need to be small in scale and run frequently if they are to connect without destroying the city they serve. Buses, of course, are already fairly small in scale compared to rail, but the trend here is for longer buses (rather than higher ones) and hybrid bus/tram technologies.

The second principle is to reclaim urban space for the pedestrian first and then public transport second. Since transfer is mostly on foot, the space needs of the pedestrian must have priority. Unfortunately, this is rarely the case and, where space is provided, there are often barriers, level changes and other impediments to smooth people flow. A good example of poor pedestrian provision through bad design is at Waterloo Station in London. Here, the connections between bus, taxi, underground and overground pedestrian bridges conspire to confuse those on foot who are intent on reaching the station. Interchange is not just about connecting transport modes within congested cities, but of putting the passenger first. Unfortunately, those who decide on the programme and draw up the design brief are the transport infrastructure providers, not the users. One simple way forward is to ensure that every interchange has an external plaza that is traffic free. This space then becomes the medium of connection, around its edges or beneath, to the new and existing transport systems.

Social sustainability and modal switch

The concept of sustainable development has social as well as environmental dimensions. Encouraging a modal switch from private to public transportation reduces dependency on fossil fuels, but significantly also helps with maintaining community cohesion. The interchange is a place in the full sense of the word – it is rich in social value, a venue for meeting as well as travelling, and a location for strengthening social and economic sustainability. The physical web of connection around the interchange becomes in time a social network and, in the right circumstances, also a channel for business regeneration. That is why it is important to plan the interchange as social as well as transport space.

Designing spaces and routes for the use of people with all levels of mobility and sense sharpness increases the standard for everybody – able-bodied and disabled. Similarly, providing high-quality materials and well-lit spaces throughout the interchange increases satisfaction levels and respect for the concept of public transport. If the attention paid to comfort and style in private cars and showrooms was extended to the public transport system, there would be greater uptake of services. Design is, therefore, not just a question of utility and function, but that of enhancing public amenity and social value through good-quality spaces, the use of attractive and durable materials, and attention to physical and psychological comfort in all the interchange areas.

Key passenger needs at interchanges are:

- safe and secure routes through the interchange;
- well-lit legible spaces and routes;
- clarity of transport mode geometries;
- clarity of signage;
- attention to disabled access and sense limitation;
- avoidance of level changes;
- high-quality materials within touching distance;
- attention to physical and psychological comfort.

3.17 Urban transport square identified by dome at Cercanias Station, Madrid. (Photo: Brian Edwards)

3.18 At Fulton Street Transit Center, New York, designed by Grimshaw, there is an interesting interplay between the interior concourse and exterior one. In both areas social interaction is encouraged through high levels of transparency and light. (Courtesy of Arup in collaboration with Grimshaw)

The geometry of spaces for people varies according to interchange type and location. However, as a general rule, the passengers require information on their point of arrival. They need to know the time and location of transport services, they need direction to facilities, and they need to know where to buy tickets. Some of this information, such as train, flight or bus times, can be placed outside the interchange, as at Copenhagen Central Station. Here, it is displayed on electronic screens above the various entry points, thereby reducing congestion at screens inside the building. The information and ticketing area is normally a distinct entry space located before the major concourses, which distribute passengers to the different transport facilities. The ticketing area is usually an

3.19 Concourse with ticket information at Waterloo International, designed by Grimshaw. Notice how the curve and height enhances the quality of the space. (Photo: Brian Edwards)

enclosed space with seats and contrasts with the more linear concourses. In the latter are likely to be found cafés, book-shops, sandwich bars and flower stalls, which line the edges of concourses. In larger interchanges their presence serves general public need as well as that of the travelling public. Often, a contained circular booking hall leads to a wide promenade-type concourse, which in turn leads passengers to platforms and gate areas. Ideally, a feature of the interchange is that the central booking hall provides tickets and travel information for all transport types and hence all users. Unfortunately, the different transport companies involved often retain their own ticket areas, making a central service difficult to deliver. With separate ticketing goes an un-unified information system for providing travel data to the traveller. Worse still, it results in one transport company not recognising the tickets of another, thereby undermining the very concept of interchange.

The key spatial pattern of a typical interchange comprises:

- a large enclosed ticketing and information area;
- an open external area for inter-modal connection;
- an internal concourse for inter-modal connection and waiting;
- a linear platform, pier or gate for boarding;
- a bridge or tunnel for crossing transport infrastructure.

Spatial planning in three dimensions

The different geometric characteristics of interchanges exist in cross-section as well as plan. Whereas omnibus and ferry interchanges exist normally on a single level, train and airport interchanges require vertical separation of the different transport systems. Although the primary mode is normally at ground level, engineering factors may make this undesirable. For example, TGV, intercity and main line train services often enter city centres at low level, thereby causing less disruption to surface activities.

3.20 Metro station within new landscaped boulevard in suburban Madrid. (Photo: Brian Edwards)

At airports the occupation of the ground level by aircraft means that other transport modes are either at high level (if light rail) or underground (if heavy rail). However, the situation is not always that simple. For example, a metro system will normally be below ground in urban centres, above ground in the suburbs and at ground level in more rural neighbourhoods. The same is often true of heavy rail transport systems. This not only allows traffic and land uses to flow above or below in uninterrupted fashion, but more importantly for the interchange, it means the cross-section holds the key to resolving movement conflicts. Design in section therefore assumes greater importance as the interchange absorbs more transport functions and becomes more urban in character. As a consequence, as much attention has to be paid to the vertical circulation systems as to the horizontal. Moreover, if changes are needed during the life of the interchange, it is often more expensive and problematic to alter the lifts, stairs and escalators than the horizontal movement corridors.

It is important that speed of transfer between transport modes is given the highest priority. Passengers do not like being delayed or being circulated unnecessarily. Speed of exit is also important – passengers need to get in and out as fast as possible. The challenge with multi-level interchanges is how to do this without compromising the efficiency of connection. The key, as Sir Nicholas Grimshaw points out, is to provide big inviting spaces that are well lit and offer views to the mode of transport and to the city beyond (Grimshaw, 2009). Such spaces also need to link well with the adjoining urban spaces in terms of both space syntax and land use function.

By giving each major mode of transport a different level, the interchange functions smoothly and legibly. Clarity of engineering design should also lead to clarity of architectural design and hence clarity for the user. The section holds the key to interconnectivity and to spatial legibility. As Leon Alberti noted in 1486, order is found in the plan of a building, but beauty (and legibility) lies in the cross-section. Designing in section has the benefit also of considering the interchange in the context

3.21 Manchester Piccadilly Station refurbished for intercity high-speed trains by BDP. Old and new stand happily alongside each other. (Photo: Brian Edwards)

of energy efficiency and the city. Matters such as height, viewing angles, sun paths and wind corridors demand exploration in section if sustainable solutions are to be found. Hence, it is important to ensure that the interchange interacts with local environmental and land use characteristics in section as well as plan, if lasting sustainable development is to be achieved.

Key factors in the three-dimensional planning of interchanges are:

- Put the most frequently used transport mode closest to ground level.
- Ensure lower concourses connect well visually with upper ones and with external streets.
- Ensure the most frequently used transport mode is the most visible.
- Avoid placing one transport mode where it becomes a barrier to another.
- Place transport systems in natural light where possible.
- Use light walls and reflectors to take natural light into subterranean areas.

- Consider the life cycle energy costs of artificial light and ventilation.
- Avoid unnecessary changes in level or direction.
- Give equal weight to legibility and route way-marking as to spatial function.
- Ensure three-dimensional space is the medium of physical and visual connection.
- Ensure speed of transfer and exit remains a top priority.

Although an exciting generation of transport interchanges is emerging across much of mainland Europe and further afield, the effects are yet to be seen in Britain. There are two main reasons for this: first, Britain has an enviable inheritance of railway stations from the past which, like St Pancras in London and Manchester Piccadilly, are readily converted to modern interchanges. Second, without investment in high-speed rail transport there is little justification for new interchange stations. With airports the situation is slightly different. Here, Gatwick, Stansted and Manchester Airports all perform, partly at least, as transport interchanges, although the plane is very much the dominant mode of travel. It is no coincidence that, in Europe, new dramatic stations have occurred where there has been large investment in TGV. Further afield the modern

3.22 Sustainable development requires effective public transportation from ferries to airports. Here the Thames Harbour bus pier, designed by The Manser Practice, provides access to an energy-efficient alternative to land-based travel. (Courtesy of The Manser Practice. Photo: Chris Gascoigne)

3.23 Welcoming entrance to Lisbon Oriente Station, designed by Santiago Calatrava. The building operates as an interchange both physically and socially. (Photo: courtesy of Torben Dahl)

interchange follows the creation of new modes of urban travel from metro systems to heavy rail. At Shanghai South Station Interchange, for example, the construction of a new underground railway and the upgrading of intercity provision led to the building of a massive interchange to accommodate transfers between rail systems and between rail and bus.

As the concept of sustainable development becomes more widely accepted, the train, tram and bus will begin to replace the car as the main means of intercity and intracity transport. After all, you can do a lot on public transport that you cannot in the car, such as work, read, eat, type on your laptop and talk on the telephone. The business case for travel by train, the sociability of public transport, stations and interchanges, and the energy efficiency of shared transportation will inevitable win over those sceptics still wedded to their cars. What is needed in the UK, however, is a renaissance in the design of public transport buildings. Put briefly, trains, planes, buses and ferries need interchanges with the same sense of style and glamour enjoyed by the car and its supporting infrastructure of showrooms and grand prix spectacles.

Taking a broad view of sustainability, it is clear that, when all energy costs are considered, the suburban model at current densities is untenable. Although a typical three- or four-bedroom suburban house is likely to be more energy efficient than a comparable older property in the town centre, this is only part of the total picture. Town centre properties are often in dense formations – perhaps as an apartment block or terrace of row houses. Here, they benefit from heat transfer between properties, thereby reducing total energy consumption. Many, too, are built above other land uses such as shops and offices, where again they benefit from thermal flows through the building fabric. Added to this, the urban microclimate is likely to be more favourable than a suburban or rural one, again reducing total energy demand. But the main difference lies in the energy costs of transport, both of people and of goods and services. Running one or two cars and journeying daily around 30 miles means that the total energy cost of suburban living is significantly higher than the urban equivalent, where the journeys are shorter and taken by public transport. Having closer proximity between living, work, education and leisure reduces traffic demand and hence fossil fuel consumption. It also increases the use and efficiency of the public transport network and the value of transport connectors in people's quality of life. Hence, the interchange does not exist on its own, but is part of a wider strategy for transforming our cities from fossil fuel dependency to more environmentally sustainable models.

What is often overlooked is that the interchange is both the gateway to public transport and, in the opposite direction, the portal through which we first view a country or city. The perception of urban place is gained by the experience of the interchange and its immediate hinterland. What view of Edinburgh, I wonder, do people gain from their arrival by train at either Haymarket or Waverley stations? It is a view of cars, ramps and traffic islands. Compare this with Lisbon Oriental or Lille Stations, with their elegant concourses and traffic-free routes into the town centre. Similarly, the Frankfurt Airport Interchange Station, designed by BRT Architekten, signals arrival into the high-tech world of sustainable transport. With its airiness, simple lines and readily comprehended routes, this airport/train/bus interchange also supports social inclusion.

Interchanges address a number of green issues, from urban space design to structural engineering. The wide spans required of concourse and platform canopies, the need to bring daylight and natural ventilation into the heart of the building, and the need to address the interchange in its varied social, civic and landscape context, all stretch the imaginative skills of architect and engineer. Equally, the booking hall and concourse capture the gathering function of the interchange; here, public life is given dignity against a background of functional complexity. There are also bridges, tunnel, ramps, lifts, escalators and stairs necessary to cross the tracks, runways, piers and lines. Under pressure to increase their retail revenue, modern interchanges are following the earlier example of airports and filling in an unsustainable way every available space with shops, cafés and vending machines. The architectural orchestration of space, route and light can be lost as people cram into the voids left over in the pursuit of retail income by transport companies. Unnecessary consumption is encouraged by extended waiting times and trapped interchange passengers. The very essence

of connectivity between transport modes, which is the justification for the concept of the sustainable interchange, can be undermined by a sense of profit rather than service. It is evident from the examples in this book that the interchange has to balance often conflicting agendas driven by the interests of different stakeholders – government that strives for sustainable development, transport companies intent on profit, infrastructure companies concerned with maintenance, retailers bent on selling products and services, and finally passengers who wish to journey. Sustainable urban transport requires the interchange to give due regard to all the social, economic and environmental dimensions of travel.

In designing modern interchanges there is much that can be learnt from examining existing railway stations and airport terminals. Over time there is often a loss of space for the passenger and also natural light. Navigation through the building is via a complex web of levels and changes in direction, with former daylight and sunlight obscured by advertising signs, shop-fronts and security barriers. Traditionally, stairs were placed in pools of light and trains and buses sat in well-lit open platforms. Today, however, the invasion of shops, stalls and advertising banners has resulted in the obscuring of natural light and, as a consequence, of views through terminals and stations. Such loss of natural light and ventilation results in the use of artificial means, which are normally delivered electronically and at high environmental cost. The grand vistas of old have been replaced by intimate short views and, as a result, there is a loss of scale, of orientation and of drama. Added to this, the pressure to maximise sales leads to a blurring of the distinction between terminal, station (and hence interchange) and shopping mall. Many existing railway stations and airport terminals are less gateways to cities than an extension to anonymous non-place retail malls, with all the unnecessary consumption and use of artificial lighting that this entails.

For the architect, the interchange has a pedigree of eco-technology whose values have endured changes in management fashion. Light, structure and detail are recurring themes capable of reinterpretation in the twenty-first century. The grand nineteenth-century railway termini of Europe and America were at the cutting edge of science and technology, and of

3.24 At Newcastle Station in the North of England, the open space of the nineteenth-century concourse has been partly filled with shops and new ticketing facilities. These often spoil the graceful lines and lofty air of the original architecture. (Photo: Brian Edwards)

sustainable design as understood at the time. Wide spans, lofty volumes, sunlit concourses and grand processional routes led to innovation in the use of materials, structural solutions, glazing and ventilation technology and in detailed assembly. Viewed from the train window, the rhythms of passing people on platforms, pools of light and unfolding city panoramas were the very essence of the station experience and arguably also of green transportation. Likewise, the tall booking halls with their circular cupolas of coloured glass provided a safe, welcoming and secure entry point to the station. As we enter the new age of the interchange we need to look back at sustainable practices in transport architecture before we jump forward.

Today's growing involvement of architects in big transport projects

As one of Arup's leading transport architects, Nille Juul-Sørensen, notes 'it is particularly important that you can see where you are going and where you have come from' (Juul-

Sørensen, 2009b). The passenger should be guided by urban and building design rather than signs, and this is one of the drivers for greater involvement of architects at the early stages of transport infrastructure provision (Juul-Sørensen, 2009a). Architects are trained to think about space as the medium of connection, both aesthetically and functionally. Hence, architects need to ensure that the transport interchange has its own language of spaces, light patterns and details of construction, which users readily comprehend as part of the bigger pattern of interchange facilities. However, unlike singular transport types such as railway stations, the interchange often has competing companies sharing a common space. The challenge is to provide route legibility in a world of potential chaos and competition for customer attention. This is why the architect has begun to challenge the supremacy of civil engineers in the initial stages of design of large transport facilities.

Engineering skills are essential, but the knowledge base of civil and structural engineering rarely engages with human values. It is satisfying legitimate human aspirations in these hothouses of transport interconnection that provides the justification for greater involvement of the design professions at the genesis of projects. This is why Arup employs 500 architects among its global staff of 10,000 – many of these employed on designing transport facilities (Juul-Sørensen, 2009b). This marks a shift in the balance of power, not just at the world's largest engineering consultancy, but in raised ambition globally for the quality of transport infrastructure. After all, public transport is one of the major drivers of urban form and requires design as well as engineering input if we are to achieve sustainable development in its broadest sense. However, the role of the architect is easier if urban design and sustainability values are put into the brief right at the start of the programme rather than added later when problems of planning consent are encountered.

The involvement of architects in these big transport projects has altered the process of design. Transport infrastructure provision used to be seen as an engineering problem, but today clients and municipalities recognise that good design can make a big difference (Grimshaw, 2009). Also, many large projects require either an architectural competition or competitive design tendering at the outset. This has raised the profile of architects and forced a reassessment of priorities. As a result, engineers and architects collaborate more today than in the past, using group working right at the start. The process is usually one of searching for hard-core facts while raising debate about design quality issues. Here, the knowledge-based tradition of the engineer and the creativity of the architect come into fruitful contact.

Design ideas start by sketching possible solutions and testing these through rough models. As Nille Juul-Sørensen (at Arup) and Sir Nicholas Grimshaw note, sketching details as well as the whole allows for quality and architectural character to be established at the start. There is a reluctance to commit too early to CAD, believing that an open exchange between architect, engineer and client is best facilitated through the sketches and models. Clients are also impressed if the architect can sketch options at meetings that not only carry the authority of the collaborating engineers but that show a commitment to passenger ideals (Juul-Sørensen, 2009b). Often too, particularly on projects outside the UK, clients like to sketch their ideas and this engagement increases the commitment to design quality.

The competition system of procurement has led to the emergence of specialist architect and engineer teams who cooperate on projects around the world. Arup is often involved as engineer but also as architect. In fact, in Arup's London office there are, at the time of writing, 90 architects engaged on transport contracts from China to Australia and the USA (Dobrovolsky, 2009). Architects who can produce clear, imaginative and realistic visions are much in demand as clients seek to distinguish their stations and airports from those of competitors. Over the past decades, due mainly to the privatisation of services, there has been a loss of in-house professional expertise within the big transport companies such as British Rail. This has encouraged the diversification of design solutions as private architectural practices are brought in as a consequence of competitive design tendering. However, one consequence of this trend is the increasing use of architects to produce the conceptual thinking, leaving the implementation to project managers or engineers (Dobrovolsky, 2009). This raises questions over the control of details and quality of construction.

Supporting sustainable development through public transport

In broad terms there are only four ways to make our cities more energy efficient and hence more environmentally sustainable. The first is to switch fuel consumption from fossil sources to renewable ones. This involves developing solar, wind, hydro, biofuels and nuclear solutions. Such fuels can, of course, power our transport system, making mobility by public transport more environmentally benign.

The second is to address the existing urban infrastructure and building stock to improve energy efficiency. This involves exploiting renewable energy sources, including geothermal, in a more comprehensive fashion. It also means upgrading worn-out and inefficient infrastructure, and in buildings improving levels of insulation, utilising passive solar technologies and addressing poor urban micro-climates. In the long run, development that is well served by modern public transport is more likely to receive the investment necessary for this upgrading than buildings elsewhere. Hence, the pattern increasingly is that of linking urban regeneration with parallel investment in sustainable transport infrastructures.

3.25 Photovoltaic panels used on the St George Ferry Terminal in New York, designed by HOK. They form a physical and energy gateway. (Photo: courtesy of Adrian Wilson)

3.26 Copenhagen's finger plan of 1947. (Drawing: Brian Edwards)

3.27a–b Being able to transfer between modes of transport is a key measure in the success of public transport. These two examples are of the train–tram connection at Rotterdam Blaak Station. (Photos: Brian Edwards)

The fourth is to plan the transport system of our cities and neighbourhoods with energy efficiency as the main driver. This would entail creating dense mixed-use neighbourhoods along and around transport hubs, positioning major facilities, such as schools and hospitals, where there are good public transport links, and providing effective interchange between transport modes, including walking and cycling. Encouraging a modal switch from private cars to public transport is crucial if current levels of mobility are to be maintained throughout the twenty-first century.

Recently, high-speed trains have begun to capture market share at the expense of planes for inter-regional journeys. However, to be effective in meeting future growth demand, interchanges need to be created where formerly single railway and bus stations existed. Integration of transport modes allows the full potential of high-speed rail to be exploited. Passengers need to be able to switch from high-speed rail to local rail, metro or bus, to cycle or taxi, and even private car, without lengthy delay or inconvenient transfer. The wider modal shift from private car to public transport is a question of cost, convenience, image and safety. All four have to be considered.

With roughly a quarter of fossil fuels being employed globally in transporting people and goods, the arena of public transport needs to be addressed with the determination surrounding the debate about energy efficiency in buildings. Low- or zero-carbon technologies can apply to buildings as well as transport, and public transport as well as private cars. The interchange occupies an important transition between these systems and allows the application of low-energy technologies in the building and in the transportation facilities that it houses. In this sense the interchange can be a gateway to sustainable living – an example of low-energy technologies applied at different scales and in different contexts.

Public transport operates on a mainly linear system. To be viable, densities above 80 dwellings per hectare are needed and ideally densities of occupation should be greater along the major transport corridors. Hence, the pattern emerges of fingers of higher-density development along rail and metro transport routes crossed by lower-density grids served by buses and connected to cycle ways. The geometry of distribution of settlements should follow the transport network with nodes at the intersection of fingers and grids. Such nodes

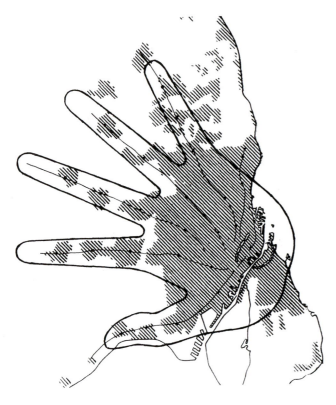

The energy and environmental strategies that are required in the design of new buildings are more demanding than those that are applied in the upgrading of existing ones. It is easier to incorporate new green technologies at the design stage than retrofit buildings later. In addressing energy efficiency in transport buildings there are three levels to consider: urban, building fabric and interior levels. Environmental design holds the third key to sustainable development and, in the case of the transport interchange, should be one of the main drivers in the design of such facilities.

will contain civic uses as well as higher-density housing. Around the interchange and within it there will be retail functions serving both passengers and the wider community. The interchange is therefore a hub of transport and civic services, and the more the number of connecting transport systems at the interchange, the greater will be the intensity of uses. As a result the interchange will be the centre of a dense, mixed-use neighbourhood hub served predominantly by public transport, cycling and walking. In this sense the interchange is a critical component of sustainable development.

This pattern of urbanism is by no means new, particularly in Europe where transport planning, land speculation and public regulation went hand in hand over the past 100 years. Development along the lines of London's suburban rail and underground system, often by the rail companies themselves, is well documented. Less well known is Copenhagen's finger plan of 1947, which envisaged medium-density development along the metro and suburban rail network with denser nodes at the intersections of orbital road and bus systems. At the

intersections, the city planners located business and community facilities for the new neighbourhoods built around interchange stations. Recently, the finger plan has been extended south-wards, with a fresh corridor of development along the new metro line extending to Copenhagen's main airport at Kastrup. At the intersection of the metro and rail system the new Ørestad Interchange has been constructed as the centre of a hotel, business and residential neighbourhood. Such develop-ment highlights the interdependency of public transport, urban planning and sustainable development.

One feature of the Danish transport system is the high level of cycle use by commuters. In 2008, 32 per cent of journeys to work were undertaken wholly or partly by bike. As a result there are extensive cycle ways in Danish cities and large areas set aside in stations for cycle storage. In Copenhagen, the typical station is an interchange between train and cycle as much as between train and metro or train and bus. Similar levels of cycle use are to be found in Southern Sweden, Holland and also in parts of the USA and Japan. The priority

3.28a Bicycle parking at Copenhagen's Bella Centre Metro Station. The lack of facilities for bike storage limits the ability of the station to act as an interchange. (Photo: Brian Edwards)

3.28b Pedestrians and cyclists mingling in the concourse of Mission Bay Station in San Francisco. Cyclists are an important and neglected element of the interchange. (Photo: Brian Edwards)

given to cycling by the city authorities in these countries means that interchanges here contain large areas of cycle storage (sometimes in multi-storey stacking systems) and wide routes for transporting cycles on to trains or through the interchange. However, since the energy costs of transport by cycle are negligible, there is a large measure of forgiveness for the inconvenience caused to other users or the additional infrastructure costs entailed.

The appeal of high-speed train over air travel depends in part upon the effectiveness of connection to other transport modes. The interchange should work like a machine and share the efficiency of the transport network as a whole. Passengers need to transfer quickly and easily from high-speed train to local services. For journeys up to 500 kilometres the train competes well with air travel and, as energy prices rise and rail technology improves, this distance will increase in the future. Facilitating effective connection between transport systems, communities and businesses helps achieve both economic and social sustainability.

Achieving the sustainable interchange is easier in the design of new facilities than in the upgrading of existing ones. Masterplanning transport provision from scratch allows the full spectrum of green design approaches to be integrated. However, with refurbishment of an earlier generation of interchanges, the task of employing new energy technologies or opening up the building to daylight is compromised by the decisions made in the past. Also, it is often difficult to thread a new mode of transport into an existing interchange in the interests of enhancing inter-modality. Without interior space to remodel, the answer here is normally that of placing the new transport mode alongside the existing interchange. This is what is proposed at the new high-speed line and station in Florence, where Foster + Partners have designed a new structure to stand alongside the existing transport facilities. An external concourse now joins old and new to form a multi-modal interchange. Although there is much disruption planned to the historically interesting Florence mainline station, without inter-modality there is little lasting sustainable transportation.

Much of Europe and parts of America and India have fine old interchanges dating from the late nineteenth and early twentieth centuries. Connected to metro, tram, underground railway and bus services, these interchanges (normally simply called railway stations) have become much cherished buildings. Many, too, are listed as structures of architectural or historic importance. They tend to contain large concourses linked to busy squares and elegant well-lit platform areas. Many also feature innovations in construction technology, spatial design or civil engineering. Adapting and extending such structures is fraught with difficulty, with the result that new transport provision is sometimes placed some way apart from existing transport structures. This is what happened at Lille Station, which operates from the town centre for national and local trains, with Lille Europe built nearly a kilometre away for the high-speed trans-Europe rail network. Although there is logic in the disassembling of parts from an engineering point of view, in terms of sustainable development (particularly its social dimension) the arrangement leaves much to be desired.

In designing urban spaces for good public transport usage, clarity and legibility of the corresponding transport interchanges is important for users, but hard to provide with underground rail systems (Hegger *et al.*, 2008: B2–3). Here, the cost of land saved by building beneath the surface is offset by the high

3.29a–c Bus–train interchange at Hamburg North, designed by Blunck and Morgen in association with Martin Tamke. The project won the Hamburg 'Building of the Year' in 2010. (Courtesy of Blunck and Morgen. Photo: Martin Tamke)

3.30 Design for Florence Station by Foster + Partners. (Courtesy of Foster + Partners)

energy costs of lighting and ventilation, let alone the social costs of loss of legibility of essential public facilities. To maximise the potential of urban interchanges, attractive paths and cycle ways and easily accessible cycle parks should be provided in urban plans (ibid.). Since, typically, every day we spend over an hour travelling, the quality of the journey has a direct bearing on our quality of life. The aim is not just to reduce the journey time and improve its pleasure level, but to encourage a switch in usage from car to rail and bus or car to bicycle. Ideally all journeys would be on foot, but the diversity of city densities and the texture of land uses rarely permits this. Instead, more effective integration of transport modes is required to make this one hour per day ideal achieve greater distance, flexibility, choice and pleasure.

One final benefit of the ever growing influence of the concept of sustainable development is the way it can potentially alter project briefs for the better. At the moment (at least in the UK) the emphasis in the programme is upon lowest price, quickest construction and hitting a number of government procurement indicators, some of which include design measures. However, there is little consideration of long-term asset value accumulation – the very things the nineteenth century left us.

Sustainable development is about long-term values; investment today, so that future generations have access to good resources (built and unbuilt). Sustainability changes the lowest-price model and puts into procurement new values and new processes.

Detailed design approaches

Having discussed broad design strategies for the interchange in earlier chapters, it is now necessary to outline a number of more detailed considerations. These will be considered under eight headings: *structural design*, *environmental design*, *spatial design*, *disabled provision*, *materials and finishes*, *lighting and signage*, *baggage and storage*, and *involving artists*.

Structural design

According to Sir Nicholas Grimshaw, the roof is possibly the most important consideration when designing transport interchanges (Grimshaw, 2009). It survives while everything underneath is in a state of flux and is critical to passenger orientation, daylight penetration and natural ventilation. Roofs need not only to float and be uncluttered, but to help define the movement corridors underneath. The design of the roof is a major concern not just in the engineering of new transport facilities, but in upgrading existing ones where commonly the architect is involved in reinstating the elegance of former roofscapes. For example, at Dresden Station in Germany, Foster + Partners invested a great deal of effort in restoring the triple-vaulted roof. Structurally, the roof often poses complex problems, especially where, as at St Pancreas in London and Leipzig Central Station, old and new roofs converge as a result of interchange expansion. As Grimshaw notes, 'at Waterloo all of our early thinking went into the roof' (ibid.). Although there are numerous other structural problems to address, it is the roof that allows architects and engineers to work together to enhance the experience of the travelling public.

Multi-level interchanges necessarily require massive structural supports to hold the weight of elevated transit systems. High-level railway lines, in particular, impose a structural rigour upon the whole interchange and often also the surrounding landscape. The construction of multi-storey railway, bus and ferry facilities poses two major problems: first, that of ordering the structural arrangement so as not to impede the reality and perception of passenger movement routes. Second, the cost rises as the number of levels increases – hence, there is pressure to share concourse space and decks between modes of transport. However, columns, beams and retaining walls provide a spatial discipline that architects can exploit to define major and minor routes through the interchange. Natural and artificial light can also be employed

4.1 Movement, structure and light are the main architectural qualities of typical train interchanges. This example is Berlin Central Station. (Photo: courtesy of Odilo Schoch)

4.2 Tectonic architectural elements at Melbourne's Southern Cross Station, designed by Grimshaw. (Photo: courtesy of John Gollings)

sometimes used, especially in Scandinavia, where it forms a link with the wider landscape both physically and culturally. However, in the new interchanges of Europe, Asia and North America, structural concrete and steel form the main framing elements. These materials in turn provide the supports for the next level of structure, which may include aluminium, masonry (brick and tile), stainless steel and timber. However, much depends upon site conditions, preferences of designers and environmental constraints. It is important, as Sir Nicholas Grimshaw points out, to concentrate on structure and detailing, since these have a timeless quality while much else is in a state of change (*Archldea*, 2009).

Because of their size and shape, transport buildings lend themselves to modularisation. The idea of modularity is one that has influenced several UK architects, from Foster at Stansted to Grimshaw at Waterloo International. Using standard components reduces cost and speeds up construction. However, site geometries mean that pure modularity is rarely achieved except at airports. More common is the adaptation of systems to accommodate the curves of railway tracks or the turning circle of buses. This results in the use of flexible component and structural systems, where the computer handles the complexity without undermining the economy of means implicit in modular construction. Curves do not compromise the concept of modularity as long as the system adopted is an open and flexible one. In many ways, the beauty of many recent transport interchanges lies in the dialogue between the discipline and order imposed by system construction and the ingenuity or imagination of the designer.

Two points are worth noting. First, the primary structure of columns can get in the way of fast-moving passengers, especially those racing to make a connection. Columns also impede movement through concourses and along platforms by those in wheelchairs, carrying baggage or pushing prams. Hence, it is better to position structural columns outside movement corridors wherever possible. However, structural support is an essential component of the interchange and, as the number of transport modes increases, so too does the complexity of the basic engineering. As a general rule, it is better to use rounded columns than square ones and to curve the corners of other constructional elements. This reduces the risk of impact damage to the interchange and injury to passengers.

to reinforce the movement hierarchies present. So, whereas columns and other structural members necessary to carry the weight of interconnecting transport infrastructure often form a barrier to free movement through the interchange, these elements can be useful architecturally.

Normally, rail systems are cheaper when placed in a cutting than on a deck. Hence, it is more common to find light rail taken at high level and buses (and sometimes traditional tram systems) at street level, all existing above the rail network. In cities, heavy rail has to pass beneath or above the road network, which often pre-exists the age of railway construction. However, the question of cross-city movement around interchanges imposes a discipline upon the whole transport system, especially as the weight of political argument and, hence, public investment shifts from road to rail. Low-level railway lines result normally in interchanges at street level, possibly with elevated light rail operating above bus, cycle and taxi levels. Railways at street level normally employ an elevated concourse where it forms a bridge over the tracks (as at Croydon near London) or a low-level one as at Leiden in Holland. Whatever the arrangement in plan and section, the structural engineering of the interchange will be a significant factor in giving the spaces their architectural character.

Steel and concrete are usually the main construction materials employed for the primary structure. Timber is

In this regard concrete (either *in situ* or pre-cast) lends itself to shaping more readily than steel, which is manufactured normally with right-angled corners.

Structural design principles are:

- Use architectural structure to define key routes, spaces and movement hierarchies.
- Employ structure and daylight together to animate spaces.
- Avoid placing columns where they obstruct movement.
- Use 'rounded' structure to avoid sharp contact.

The second point concerns the environmental role of the primary structure in tempering the interchange climate. In both hot and cold countries, the thermal capacity of concrete can be utilised to moderate the interior climate of booking halls and platform areas to the benefit of passengers. Although steel is useful in providing support for large glazed areas (which in turn serve to enhance the comfort of passengers), one function of material selection by the architect is that of achieving energy efficiency and climate moderation by natural means. So, whereas structural considerations will necessarily be to the fore, it is imperative that the environmental and ecological dimensions are also addressed.

There is a reciprocal relationship between the frame and the nature of the spaces created. Certain structural systems lend themselves to the enclosure of round spaces, others to linear spaces. The language of structure and construction can help define the nature and function of the different interchange volumes, thereby aiding legibility for the user. Such legibility extends to the three main functional elements of the typical interchange: large enclosed volumes for booking and information, linear spaces for boarding, and movement zones for joining up the different transport modes present. If these are treated architecturally in different ways using structure as the defining characteristic, users will be better able to navigate their way through complex interchanges.

As already mentioned, architectural structure is a useful guide to directional orientation. People learn to read the routes through complex interchanges by the spacing and size of columns. The structural frame and its secondary elements serve at least two functions: they provide support in a practical sense and they support perception of the routes and hierarchies present. Hence, structure is often more expressive than needed functionally in order to compete with all the other elements that make up the internal (and external) environment of the interchange.

4.3 Platform structure and lighting at Atocha Station, Madrid, designed by Rafael Moneo. (Photo: Brian Edwards)

It helps with spatial perception and route legibility if the strategy for architectural structure is combined with that of natural lighting during the day and artificial lighting at night. Bathing the primary structural members in daylight helps signal their importance and is particularly helpful for people with visual impairment. However, it is vital that sunlight is controlled in order to prevent glare, particularly around staircases and where columns stand within major routes through the interchange. As a general rule, major structural members should signal the significant movement corridors and concourse spaces with daylight reinforcing the message. Rather like a city, the major routes are the civic streets of the interchange and the booking hall and concourse areas its squares. Using the language of tectonic architecture helps communicate the patterns of functions present.

Travel buildings tend to be long-lived. Typically, the primary structure survives for at least 100 years, with cladding, glazing and other space enclosure replaced more frequently. One role of engineering and architectural design is to allow the different structural and non-structural elements to be upgraded without disrupting the whole. If the main engineering of the interchange remains relatively unaltered, this is not the case for the transport modes present, which typically have a life of 20–30 years at best. Their upgrading and the new transport technologies that support them stress existing space patterns as well as the architectural structure. There are also frequent changes to management practice, in the ownership of retail interests, in security and ticketing policy, and to legislation on topics from disabled access to health and safety at work. Collectively, it results in little long-term stability for the initial design concept. How transport buildings adapt varies between types, but invariably it is the amenity for passengers that suffers. Alteration means less movement space temporarily, loss of legibility and changes in familiar space patterns. Creating flexibility at the outset should be the aim, but few design briefs put long-term change into the procurement documentation or allow for the extra costs of building long-life, loose-fit, low-energy transport interchanges.

Upgrading cycles for typical transport interchange are:

- main structure: 100 years;
- external walls: 50–100 years;
- glazing and cladding: 25 years;
- environmental systems: 20 years;

4.4 Flintholm Interchange north of Copenhagen exploits solar gain to provide a comfortable place to move between rail, metro and bus services. (Photo: Brian Edwards)

- internal walls: 15 years;
- carpets, seats, lights: 5–10 years.

Environmental design

Considering the important role public transport plays in supporting sustainable development, there is a surprising lack of examples of transport interchanges that demonstrate the principles of sustainable architecture. Admittedly, a number of recent stations in Holland, Denmark and Germany have sought to maximise solar gain for background heating and stack-effect ventilation, several airports have introduced natural light and ventilation in deeply planned terminals, and Norway has built the surprisingly green Gardermoen Airport in Oslo. The latter is an interchange of some distinction and employs locally sourced timber and slate, which, when combined with high levels of daylight and solar-assisted ventilation, begins to balance the equation of energy-efficient transport and ecological design of the building. The linking stations between Oslo and the new airport interchange engage in a similar ambition to be examples of sustainability in a broad sense for the communities served.

These and other examples highlight a trend in recent green transport architecture practice, evident at Grimshaw's Fulton Street Transit Center in New York and elsewhere, of combining local heavyweight construction materials with imported lightweight ones. The former provide regional references that help with cultural acceptance of the interchange, while the latter bring to the building the best of high-tech energy-design practice. This high- and low-tech combination occurs in many interchanges and allows for sustainability to be interpreted differently in different regions of the world. Besides Grimshaw's New York example, one could cite Madrid Barajas Airport or Frankfurt Airport interchange or Yokohama Ferry Terminal. What they demonstrate is the interest among architects to anchor their buildings in specific places with specific climates rather than apply a recipe irrespective of context.

The sense of place specific-ness is a reflection of the growing interest in cultural sustainability. To users of transport facilities, the nuances of culture matter in an age of rapid industrialisation (as in China). Some architects walk the streets before they commit to design, in order to gauge those subtle interactions between culture, climate and building tradition that still exist in the world (Juul-Sørensen, 2009b). It also helps to

4.5 Sculptural forms used at Lisbon Oriente Station, designed by Santiago Calatrava. The architecture creates attractive patterns of light and shade. (Photo: courtesy of Torben Dahl)

see what forms of construction are employed in new buildings and, hence, how traditional values can be incorporated into the high-tech world of transport architecture. Often the solution consists of blending modern and traditional values, as in the transport buildings in South Korea by Sir Terry Farrell, in China and Germany by Lord Foster and in Spain and Portugal by Santiago Calatrava. New transport architecture can help maintain the differences in the world rather than erode them into a homogenised whole.

Like all buildings, transport interchanges consume a great deal of energy in operation, have large environmental impacts during construction, and have tentacles of ecological impact through the adjoining neighbourhood. However, unlike many large civic buildings, transport interchanges are not normally heated throughout. The open nature of this building type, with its multiple entrances and canopied platform areas, results in less energy consumption per square metre than for an enclosed building. Normally only those parts of the interchange where people sit, gather or work are heated. Typically, the booking hall is heated while the platform areas are not. However, there are exceptions. At airport interchanges it is common practice to heat concourse areas, gate lounges, access piers and linking movement spaces. At the interface between bus and

plane or between train and plane, the passenger sometimes moves between heated and unheated volumes depending upon the policy of the transport mode operator. For the user this can appear irrational and disarming. Another exception is to find heated waiting areas on unheated platforms and concourses. So, rather than have a single strategy for heating the whole interchange to an even temperature, designers and engineers normally specify a temperature range for different zones. As a general rule, it is assumed that in the spaces where people sit, work or wait for long periods there is the need for heating to normal room temperature; in malls and corridors ambient heat is sufficient; and in open access areas (such as rail or bus station platforms) no heating is needed. However, much depends upon the background climate and local cultural traditions.

Since interchanges contain large volumes exposed to sun and wind, much of the energy required for heating and ventilation can be gathered from renewable sources. The sun provides two obvious benefits for sustainability: it can be employed for passive solar heating of both concourse and platform areas, and it can provide electricity using photovoltaic technologies. Passive heating also leads to passive cooling using the thermal dynamics of the different interchange spaces.

4.6 Barnsley Bus Interchange in Yorkshire, designed by Jefferson Sheard, provides a high-quality environment by maximising daylight and sunlight in the concourse spaces. (Photo: Brian Edwards)

Similarly, wind-assisted ventilation can be utilised, perhaps augmented by passive solar cooling.

Water can also be harvested from the large roof areas and paved terraces of a typical interchange. Such water could be used for cleaning platforms, concourses and staircases, for irrigation of landscaping, for fabric cooling and for toilet flushing. Rainwater should not be viewed as a problem of sewage or waste disposal, but as a potential asset to the building. Collecting, storing and using rainwater reduces the energy demand and cost of piped supplies and helps saves potable water for more essential purposes.

In hot countries the strategy for water harvesting, heating and cooling will be different from that in cold climates. However, in both situations using sun and wind to reduce dependence upon fossil fuels carries enormous benefit in an age of global warming. Since the climate is variable both during seasons and through the day, the interchange should be able to respond to different situations. Louvres, shading and the ability to open sections of wall and roof are desirable features, as long as the technology is kept relatively simple and is designed in a vandal-proof fashion. It makes a lot of sense, as at Shanghai South Interchange, if the strategies for water conservation and energy reduction are tackled as integrated problems.

Much of the energy used in a typical interchange is for artificial lighting and mechanical ventilation. In some climates, the cooling load is greater than the heating one, in others the background heating from lighting (especially where large shopping areas are present) provides all the space heating. Lighting consumes a great deal of electrical energy that, per unit of power, is more damaging in terms of global warming than energy directly derived from fossil fuels. Hence, it is imperative that natural light is utilised wherever possible. It should be taken into subterranean areas of the interchange, be employed in tunnels and along internal malls, and it should not be blocked by invading shops or advertising banners. The use of daylight is critical to energy consumption and to the ability to read the visual language of the interchange. Hence, there are social as well as environmental arguments in favour of more natural conditions.

Environmental design principles are:

- Maximise natural light.
- Maximise natural ventilation.
- Use passive solar for heating, cooling and ventilation.
- Avoid glare by shading direct sunlight.
- Utilise renewable energy including geothermal sources.

4.7 Traditional cast-iron and glass roof at the concourse of Liverpool Lime Street Station. (Photo: Brian Edwards)

4.8 Modern concrete and glass roof over platforms of Oslo Station. (Photo: Brian Edwards)

4.9 Space for people to wait with information nearby is essential at busy interchanges. This example is Shanghai South Station designed by AREP. (Photo: AREP/T. Chapius)

- Use roofs to capture rainwater.
- Source materials locally.

Linked to questions of solar and daylighting design is that of natural versus mechanical ventilation. Most large interchanges operate in a mixed-mode fashion. In some parts there may be air-conditioning (in say office areas), in others mechanical ventilation (in concourses and booking halls), and in others merely natural cross-ventilation. However, the fumes emanating from the trains, planes and buses may present local problems, particularly in areas below ground. Placing platforms in tunnels or deep cuttings poses a particular difficulty for ventilation and also for fire engineering. Hence, if underground travel is unavoidable, the costs of operation and maintenance add to already considerable construction costs.

There is, though, one potential benefit of subterranean systems. It may be possible to utilise geothermal sources for heating or for cooling. Using ground source heat pump (GSHP) technology, it may be possible to heat underground platforms in cool climates or, working in reverse, to cool those in hot ones. Since temperatures underground are relatively stable, the ground source energy can act as a heat reservoir in the winter and a heat sink in the summer. If the potential of geothermal energy was linked to other parts of the interchange, it would be possible with intelligent engineering to augment the heating and cooling system elsewhere with considerable energy savings overall. Cooling is more common than heating in underground systems. This is partly because of the low levels of ventilation, but also as a result of heat given off by the transformers that are located at low level in many metro systems. Using ground water for cooling is rare, but GSHPs offer a potential technology for the future.

The three major environmental considerations of heating, lighting and ventilation need to be considered as related and interdependent problems, where the building services strategies would vary from one location to another and from one interchange type to another. Since solar and wind opportunities have different characteristics geographically, designers should follow the principle of seeking to maximise renewable energy opportunities in whatever form they take. Sometimes, also, traditional practices in the neighbourhood point to innovative use of energy and construction methods that could be applied at the interchange. Such use of an adapted technology from local vernacular practice would help anchor the new structure into the cultural landscape that it serves.

Spatial design

As mentioned already, there are three types of space at a typical interchange. There is gathering space where travel information and tickets are obtained; there is concourse space where promenades and shops occur; and linked to the concourse there are the linear spaces (such as platforms) that provide direct access to means of transport. In addition to these primary spaces, there are usually staircases, tunnels, bridges, ramps, lifts and escalators. Normally, with an interchange there are also external gathering or meeting spaces that serve both a social purpose and as the means of linking to other transport modes (buses, taxis, cars). The different types of space have their physical characteristics: some are rounded and contained, others linear and open, some can be lit naturally, while others depend upon artificial light and ventilation.

Use space design to distinguish priorities:

- travel and boarding;
- information;
- meeting;
- eating, reading, waiting.

Interchanges work as a machine and the role of space is to lubricate the movement of people between the different modes of transport. In this sense, space is for people. However, space has to work functionally and psychologically. It should be possible for travellers using the interchange to read the spaces: their function, hierarchy and meaning.

Legibility of the interchange is more important than the pure matter of time to reach any point – hence the perception of movement matters more than the functional efficiency of getting from one part to another. Spatial design, therefore, is a matter of ensuring that there are internal landmarks and guides to allow for the reading of the spaces, routes and volumes. Such landmarks may be internal, such as a clock or public art, or they may be external, such as the opening up of a vista of a

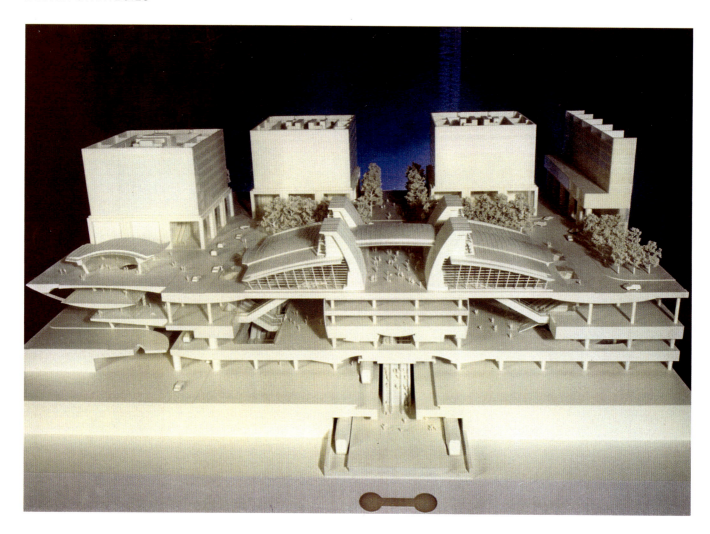

city spire. Effects of light are also important if used in an artistic fashion – the use of coloured glass is particularly successful in helping people to navigate their way through complex terminals.

With an ageing population (by 2020 it is estimated that nearly 50 per cent of Europeans will be over 50), attention to the psychology of spaces is vital if people are to be helped through the maze of a typical inner-city transport terminal. The perception of spaces is not just based on architectural features such as materials, light and structure, but includes social references such as how busy areas are, the level of noise and the position of tranquil corners with seats and planting. Having places to sit and watch the interchange processes unfold helps with gaining a grasp of the overall picture. Also, perception helps if there is a clear sense of direction, such as that of moving towards the desired mode of transport or in the opposite direction towards the town. Hence, visual clues are as important as making functional references to the different spaces.

Reinforcing legibility through design is essential if passengers are not to become disorientated. The larger the transport building the more important is the question of waymarking. Big interchanges can be seen not as single buildings, but as small cities. In his book *The Image of the City* Kevin Lynch identifies

five characteristics of the perception by the public of urban imagery. Each, he argues, should also inform designers of places such as city planners. A parallel can be drawn between city planning and transport planning. Whereas Lynch identifies the five characteristics at an urban scale, it is argued here that architects faced with the design of interchanges should follow similar rules. Applying the rules of *The Image of the City* to help create an imageable interchange, means translating *district*, *landmark*, *node*, *path* and *edge* into elements at a smaller scale.

The *district* is the whole interchange with its own transport character, and its own distinctive typologies and functional parts. The sense that the interchange has its own personality and distinctive forms is important to the identification of use. Transport infrastructure imposes a character to the whole, but so too do the people movements, which should not be denied. The weaving of transport and people is the basis of the transport district.

The *landmark* may be the interchange itself (as with Grimshaw's Melbourne Southern Cross Station), or there may be a new structure alongside an existing transport landmark (as at St Pancras Station in London). Landmark status should be reserved for public architecture and the interchange should,

4.10 At Kowloon Station, designed by Terry Farrell and Partners, the functional hierarchies are expressed by different architectural volumes. (Courtesy of Terry Farrell and Partners)

4.11 Concourse at St Pancras Station. This is a 'node' in social terms. (Photo: Brian Edwards)

4.12 Urban tram interchanges are civic nodes as well as meeting places, as here in Zurich. (Photo: Brian Edwards)

therefore, aspire to be a landmark. Zaha Hadid has followed this approach at the Strasbourg Bus Interchange, where the zigzag plan and dramatic roof cantilever signal the presence not of background architecture but of civic ambition. Being a landmark helps with spotting the interchange in the busy urban scene. Exploiting landmarks from within the interchange can also help orientate users of the building. Opening up views of city landmarks gives a sense of direction within disorientating buildings and their perimeter spaces.

Lynch describes the *node* as a place entered into where human or commercial use intensifies. One type of node identified by Lynch is the transport node. To maximise the image qualities of the interchange it is important, Lynch says,

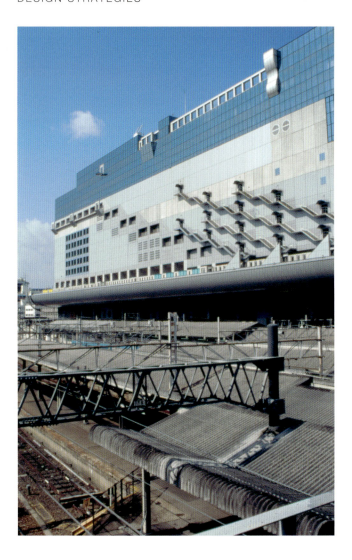

4.13 Kyoto Railway Station and adjacent hotel. The hotel and railway offices seek to bridge the tracks. (Photo: Brian Edwards)

4.14 Unless adequate facilities are provided for passengers they will take over concourse spaces. St Pancras Interchange, London. (Photo: Brian Edwards)

the interchange routes are memorable. This may entail installing public art (as at Copenhagen Airport) or creating dramatic interior routes (as at St George's Ferry Terminal in New York), exciting exterior ones (as at Yokohama Ferry Terminal) or landmarking the crossing of transport routes (as at Vauxhall Cross Bus Interchange in London). Without memorable routes there is little pleasure in using interchanges.

The final quality is that of *edges.* These, according to Lynch, are the boundaries of perception and are often formed by transport barriers of one form or another. Typically, an edge is a railway line that cannot be crossed or a busy urban road. Interchanges are often surrounded by edges that form boundaries both physically and perceptually. One role of the interchange is to bridge the boundary condition, to allow perception to cross a physical barrier. This can be done by bridges that span transport infrastructure (less so tunnels), providing not only connection but views of the web of transport provision. Sometimes whole interchanges are built as bridges, which allow passengers to drop down to the transport facilities below, but also allow others to use the building to cross a former urban barrier. At interchanges edges should be eroded and permeable.

Space design principles are:

- Ensure the interchange works like a machine.
- Ensure there is legibility of key routes, spaces and transport modes.
- Use *The Image of the City* to create memorable spaces.
- Ensure queues do not block concourses.
- Link interior and exterior spaces.
- Pay attention to the periphery of the interchange.
- Use landmarks to orientate passengers.
- Provide tranquil as well as busy spaces.
- Provide disciplined zones for retail.
- Articulate clearly information zones, movement spaces, retail zones and waiting spaces.

The larger the interchange, the greater will be the pressure from retailing on the interior spaces. Shops, bars and cafés serve a useful function and generate income that can be used to help maintain the public services on offer. However, retailing can complicate the pattern of routes and congest concourses.

to increase the level of economic or social interaction. A node is only imageable when there is engagement between space, people and activities. The more the interaction, particularly socially, the more the interchange becomes memorable. A node is also a 'place' in perception terms – it is not abstract space or just territory for economic encounter. The problem with many interchanges is that they are too functional, too machine-like in their rational planning, and too dominated by commerce. A transport node is foremost a place for people.

Lynch attaches great importance to *paths*. These are the routes taken on journeys either on foot, by public transport or by car. For the interchange the routes and paths are especially important to perception. There are the routes to and through the interchange on foot, the different perceptions gained while using other forms of transport, and the interaction between the two. We gain a knowledge of the spatial geometry of places via the routes we take and the speed at which we take them. At walking pace much detail is absorbed, even more while waiting, while on public transport the greater speed gives a coarser grain of perception. As the routes are our main vehicles for gaining the tools for navigation, it is important to ensure that

This may confuse the elderly or slow down those trying to make last-minute connections. So, whereas commerce can add sparkle and colour to the interchange and help diversify sources of income, shops and bars need to be kept in well-disciplined zones to avoid conflict with more essential functions. At large transport hubs retailing may provide nearly a quarter of total revenue for the interchange provider, and more at airports. However, although cafés may provide services such as toilets, not all travellers wish to have their desire lines through concourses interrupted by encroaching tables and chairs.

Aligning the needs of retailers is often at the heart of interchange design (Dobrovolsky, 2009). Commerce is an important source of income and many large stations operate as supermarkets for their local community. Achieving alignment between retail and transport interests can be difficult, especially as the scale and complexity of commerce at interchanges increase over time. Rather than fill the concourse spaces with shops, bars and market stalls, it is better to design the retail areas at the outset, using management agreements to curtail the colonisation of travel or waiting areas. This is why stakeholder cooperation is necessary at the inception of transport projects.

There is an inherent conflict, however, between interchange operators and retailers. The main task of the interchange is to provide efficient and pleasant connections for those travelling, while the retailer wants to slow down movement in order to get people to purchase goods of various kinds. Slowing down passenger movement and making people wait for periods of time leads to impulse buying, which is just what shops and cafés want. Cynically, it could be argued that the excessive check-in times at many airports are not to aid security or airline efficiency but merely to maximise retail sales. Placing the retail product or café snack within the visible radius of passing passengers is what commercial interests want, while the interchange operator and user prefer to have views of travel information and their desired travel mode. Clarity of way-finding needs to come before commerce if the interchange is to attract users of all ages and income brackets.

Some argue that commercial zones in large travel concourses invigorate the general environment for the benefit of all. Experience suggests quite the contrary. While many do benefit from the seats provided by cafés, these are often provided at the expense of general passenger seating. Often, to gain a seat, the passenger has to buy a cup of coffee, usually at an elevated price. For many older people this price is too high to pay, with the result that many simply stand. Equally frustrating is the provision of toilets only within the commercial outlets. In some interchanges passengers have to

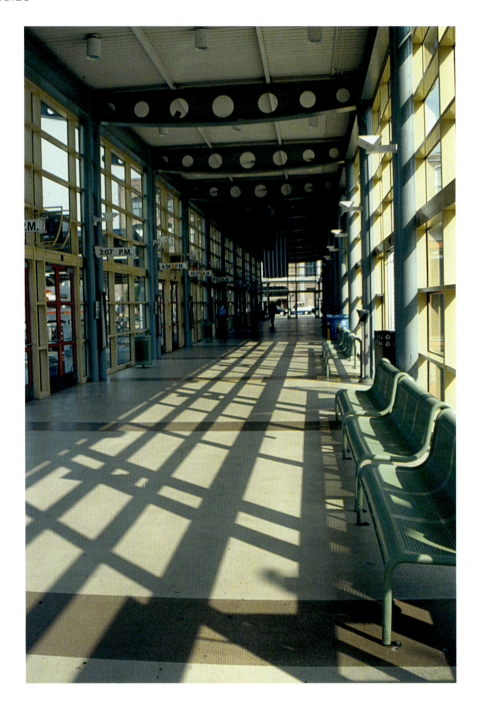

purchase a drink to use the only 'public' toilet available and this is sometimes down a flight of stairs where it is inaccessible for disabled users.

Since the interchange is a place of travel connection, it is also somewhere where journeys occasionally disconnect. Hence, there is demand for seats, tables, telephones and toilets. These need to be provided even if there is no obvious revenue benefit. The social dimension to public transport and, increasingly, its role in business travel require seats for the elderly, tables for laptop users, internet cafés and points, payphones for those without a mobile connection, and toilets for all.

Although this is generally true of all transport types, underground metro or city rail systems lack the social life character found above ground. Being below ground removes the access to daylight and view that is essential to 'coffee culture'. Cafés exist best where they can spread outwards on to pavements or into sunlit concourses. These do not exist underground and, hence, subterranean stations or interchanges are functional places where the speedy movement of people and the effective boarding of trains matter most.

The larger the interchange, the more varied will be the functions and corresponding spaces within the building.

4.15 Seating at Mission Bay Station, San Francisco. (Photo: Brian Edwards)

4.16 Detail of seat at Lille Europe TGV Station. (Photo: Brian Edwards)

Besides retailing, there may be small police offices, security areas and tourist information points within the interchange and possibly civic functions such as library, council offices, school or health clinic in the immediate area. From a sustainable development point of view it makes sense to position commercial and civic activities around transport nodes. The more the interchange is inter-modal, the more it becomes easily accessed for a wide range of people. If the surrounding area is relatively densely populated, there will be sufficient people within walking or cycling distance to sustain a significant range of services within and around the interchange. However, the reality is that the spaces around interchanges are often poorly designed for pedestrian needs. They can be windswept and bleak, with barriers to movement on foot, or civic spaces that are hard to read and frustrating to use. Even when the physical nature of the routes is clear, they often suffer from poor lighting, shoddy finishes and graffiti. This leads to a poor image of the interchange, which investment in the interior environment does not overcome.

Landscape design can hold the key to improving the external image of the interchange by creating leafy boulevards and well-designed squares that link it with the town it serves. Shade created by planting around the interchange makes the approaches attractive to use and comfortable to wait in. After all, the interchange is a place where waiting and meeting occurs. Too often this is conducted in windy, cold, open and unappealing environs just at the edge of the building. There are two particular problems that need to be addressed. First is that of bicycle storage. As energy prices rise and cycling becomes more popular for commuting, the parking of large numbers of cycles at transport interchanges threatens to restrict movement by other users, particularly those on foot. Unless vertical or compact stacking is employed, bicycles left for the day can obstruct pavements and other routes around the building. This problem is exacerbated when bikes are parked in racks at right angles or 45 degrees to external walls. Many thousands of bikes are parked daily at larger European interchanges, not all of them in official cycle storage areas.

The second 'edge' problem that defaces many transport buildings is that of smokers and their litter. Since smoking is

4.17 Relationship between commerce and movement spaces at Berlin Central Station. (Photo: courtesy of Odilo Schoch)

not generally permitted within interchanges, those needing a cigarette congregate at public entrances, obstructing mobility for other users, polluting the air, and littering the ground with their butts and packets. When drinkers also gather in these areas right at the periphery of the interchange, there is the added problem of beer cans and anti-social behaviour. Often, too, taxis wait at interchange entrances, usually with their engines running, adding further to pollution. Urban, landscape and building design needs collectively to tackle these problems, which mar many otherwise successful transport buildings.

Many homeless people use the shelter and anonymity of transport buildings for casual overnight sleeping. Often it is condoned by public authorities aware of the plight of many homeless people in big urban conurbations. It is difficult to cater for this group in design terms, yet to deter them by security barriers and policing may subject many young and insecure people to worse conditions elsewhere. Since many homeless people are drawn to interchanges by the heat and shelter provided, these buildings have also become places where social workers conduct their business. As a consequence, large transport buildings, including interchanges, are where free meals are distributed, often in discreet areas away from public gaze. Access here for vehicles providing hot soup, blankets and welfare support raises further difficulty for designers of the public spaces around such buildings, who are keen to maintain traffic-free conditions and high levels of street lighting.

The loss of route legibility or convenience of movement between inside and out, from these and other causes, results in transport interchanges that fail to interface successfully between the public and private realms. Space that is the medium of connection needs to be contained by architectural means, animated by activity and articulated through attention to detail. Low density of development outside the station fails to do either of these things. Smart technologies within the interchange should extend their presence to the peripheral areas, thereby blending the transport infrastructure with its urban context. The innovative infrastructure of modern transport planning should lead to smart new approaches to urban design.

4.18 Welcoming entrance to Manchester Piccadilly, from bus stands. Station designed by BDP. (Photo: Brian Edwards)

It is important to ensure through landscape and through urban and transport design that the interchange provides for a wide range of social interactions. People exchange journeys at interchanges, but they also engage in wider social, economic and cultural interactions, many of which generate income, social capital or cultural discourses. Typical is the buying of groceries or newspapers, listening to music played by itinerant musicians in travel concourses, or the sharing of ideas as one waits in line at booking halls or on platforms. Design in its broad sense should promote rather than hinder such engagement.

One difficulty with exploiting the air rights above an interchange for commercial or social development is the potential loss of daylight and ventilation opportunities within the building itself. Large concourse spaces and transport squares need the sparkle of sunlight to keep them attractive throughout the day. Overdevelopment, though it has advantages else-where, can be at the expense not only of energy efficiency, since daylight optimisation is critical, but also of public amenity. Losing the chance to utilise solar gain for space heating or natural stack-effect ventilation because of commercial pressure is frequently encountered in urban situations.

Increasing densities in plan and section at interchanges helps identify their presence in the urban scene and raises

expectations of a better sequence of civic spaces. The new transport interchange is a commercial and community hub where social provision goes hand in hand with economic exploitation. Mixed-use facilities, especially for those not dependent upon private car access, means that the interchange will contribute towards social inclusion and community well-being. The aim is to improve interaction between modes of transport, types of development and, as a consequence, different groups of people. This was the ambition behind the St Pancras Eurostar Interchange in London and the redevelopment of Kyoto Station in Japan. In both cases, new urban malls and squares have been constructed to support access to the new commercial facilities while also enhancing the wider transport landscape.

It is important that the spaces inside the transport terminal link smoothly with those at its periphery, with major internal concourses angled to be a continuation of external streets. Squares inside the interchange should connect directly with squares outside, and these in turn should lead to streets, walkways and cycle ways that extend like access ribbons (I am indebted to Fiona Scott's Ph.D. thesis from RCA for the term 'access ribbons' in this context) into the wider community. The emphasis should be upon public transport connectors, not private cars or taxis, with a clear policy for pedestrian priority

along the major town routes serving the interchange. Investing in people first and vehicles last means the spaces created will be of a high standard in terms of finishes beneath the feet and shade above the head. Hence, landscape design will need to integrate with urban design to create a network of green shaded and sheltered routes to the interchange.

The interchange is a place where people exchange one type of transport for another. It is a place and not necessarily a single building. Hence, spaces for exchange are likely to be internal, external and above or below ground. For exchange to work well, spaces need to be legible, safe, secure and accommodating of the different volumes and times of flow involved. Since the interchange is also a place in the civic sense, it is likely to be (or become) the location of commercial and institutional functions. So the task of the designer is to convert the notion of abstract space into places that carry meaning for the user. The phenomenology of place is quite different from the mechanistic perception of space that an architectural brief might otherwise suggest. Nowhere is this conversion more difficult than in the arena of transport design.

Large concourse spaces can pose acoustic problems. The mixture of hard surfaces, big volumes and much background noise (such as travel announcements) results in travel spaces that have high reverberation times. In order to reduce the echo time, it may be necessary to introduce acoustic panelling at high level or employ more soft surfaces generally. However, the more the concourse spaces are occupied by people, the less the problem of overextended reverberation time, since people act as sound absorbers and deflectors, but, at the same time, the greater the level of background noise. So particular attention is needed to ensure that the acoustic environment is well balanced and that pockets of noise are eliminated through design measures as well as by attention to noise attenuation through the selection of surface materials.

The notion of *form follows function* raises the question of what exactly is the function of a transport interchange. Mention has already been made of its primary role of connecting people to transport modes of many types. Movement is the key characteristic of transport nodes and, hence, the principal function of interchanges (of whatever type) is to provide spaces

4.19 Merging of streets and platform at Stadelhofen Station, Zurich, designed by Santiago Calatrava. (Photo: Brian Edwards)

4.20 Development of the air rights over Manchester Piccadilly Station. (Photo: Brian Edwards)

for movement and for travel information. In this sense *form follows mobility*, making architectural spaces fluid and dynamic rather than solid and static. There are, of course, limits to the degree of spatial fluidity, but the elements that interfere with movement (such as shops) are not serving the interests of the primary function of transport interchanges.

There are many stakeholders who help fashion the design and use of space at interchanges. Besides the infrastructure company, which may or may not own the travel buildings itself, there will be the different transport companies present (bus, train, ferry, plane), there will also be the retail interests and there will possibly be police, customs and other controls. One task of spatial design is to ensure that these stakeholders are not operating in conflict with each other. A typical conflict of interest common to many transport buildings is that of cafés spilling into concourse areas and thereby interrupting the flow of passenger movement. Another is that of security, with its barriers across desire lines. The architect cannot always predict the changes to the use of space that occur, but clear demarcation of zones for different interests helps with the smooth operation of the interchange.

Generally speaking, space is a desirable quality in transport interchanges. It allows the various modes of transport to be seen, it provides sufficient areas for movement, it enhances the sense of the interchange experience and it provides a calming effect (Blow, 2005: 18–19). When lighting (artificial and natural) subscribes to the same attention to scale and detail as that of space design, the effect can be uplifting, even in the busiest of transport buildings. Spaciousness is important, but too much space for too little usage can be intimidating. Another consideration is that of air quality. Space, light and good air (not air-conditioned, perfumed or recycled) is essential, even in underground facilities.

Transport interchanges that contain large volumes beneath the surface pose particular problems. Besides the issues already mentioned, there is the added problem of how to shape the spaces in architectural terms. Above ground, concourse spaces and platform areas are dictated by external connections, the logic of infrastructure layout and the demands of architectural structure. These also exist underground, but it is rather more a case of carving out volumes from the local geology. Here, the architect is sculpting out rather than

articulating rooms whose form and dimensions are dictated by structure and construction. Hence, beneath ground the architect is modelling out of solids rather than creating forms in free space. The emphasis on underground transport systems is that of tunnelling and excavating, creating space and light where previously all was solid matter.

Subterranean urbanism requires particular attention to be paid to design by section to ensure that the volumes created by carving and tunnelling have quality as negative forms (Spurr, 2009). They are, in this sense, the inverse of the positive volumes created above ground. However, since light only enters from above, the architect has to exercise ingenuity to ensure that it penetrates to the lowest levels.

The role of space design is:

- to differentiate stakeholder and functional interests:

    1. travel operating companies;
    2. infrastructure companies;
    3. retailers;
    4. security;
    5. passengers;

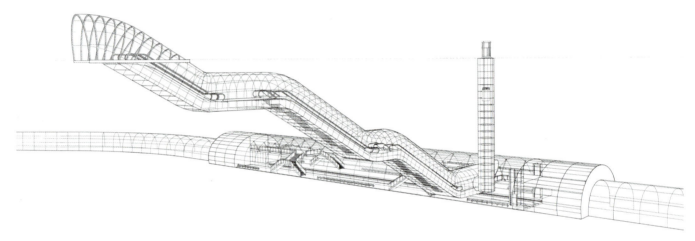

4.21 In the redevelopment of Manchester Piccadilly Station, sound-absorbing panels and ETFE were used in the roof to improve acoustic conditions. (Photo: Brian Edwards)

4.22a–b Shaping the volumes underground and connecting them to the surface is a particular challenge when designing metro systems. (Courtesy of Foster + Partners)

- to assist passengers and their comfort:
  1. to provide calm areas for rest or waiting;
  2. to support way-finding;
  3. to provide sightlines to travel modes and exits;

- to mirror functional hierarchies:
  1. to use space to mirror functional flows;
  2. to create character in different spaces;
  3. to provide clear geometries of movement;

- to facilitate change:
  1. functional change;
  2. management change;
  3. operational change;

- to celebrate public transport and inclusivity:
  1. architectural space for all;
  2. processional routes that are barrier free;
  3. quality of design that endures.

Disabled provision

The question of disabled access is an important aspect of space planning and more detailed design. There are three main areas that should be addressed in order to overcome the barriers that may limit the use of public transport by disabled people, the elderly and those with small children. As mentioned earlier, these are the very groups that the interchange should be supporting, yet frequently both design and management practices act as disincentives to wider use.

The first consists of physical measures that enhance access by both foot and wheelchair. There should be steps and ramps at every change of level. It is easier for wheeled vehicles, such as taxis and buses, to climb an incline than a wheelchair or baby buggy. Hence, entrances for people should be flat, with platform ramps at road crossings where kerbs occur. Ramps and stairs should also be placed in obvious positions and be well lit, and preferably overlooked by other activities to provide surveillance. Lifts, which are mandatory in many countries in order to cater for disabled access, and escalators, which are

also needed within larger interchanges, should be located where they can easily be found (not tucked away in corners or behind retail stores). Handrails and stair treads are sometimes illuminated to aid users with poor vision and this has the effect also of drawing attention to their position in busy locations. In colder climates handrails are also heated, partly as a by-product of the lighting element within them.

Distance, too, is a barrier for many and, when walking lengths exceed 200 metres, consideration should be given to providing moving pavements or electric people carriers. Policy varies in different regions of the world but, generally, those interchanges attached to airports will have greater travel distance on foot than other interchanges. Likewise, at large city rail centre interchanges, walking distances can be high (as at London's St Pancras Eurostar Interchange) and may require travelators or the provision of electric people carriers within the concourse. The decision on whether to provide such facilities depends not only on walking distance, but on the likely level of use by disabled people and whether there is a high percentage of passengers with baggage.

The second barrier to use is the lack of toilet facilities. Young mothers and disabled people require their own toilets and other facilities. Some interchanges have no public toilets at all and rely upon those in cafés or pubs within the building, or assume people will use the facilities on trains or buses. This is unacceptable to many older people and is not appropriate for young children. Public toilets are increasingly rare in many towns and equally scarce in some public transport buildings. The effect is to deter use or remove the pleasure of journeying at all.

The third barrier is that of perception. Many older people fear crime and finding that there are no facilities such as toilets when needed. The interchange environment should be a reassuring one where there are clear direction routes, clear signing of toilets, and places to sit and rest, where lighting illuminates the stairs rather than the shops, and where travel information is clearly displayed. The latter requires attention to font size, illumination levels and to the position of travel signs. Perception is overcome by providing the facilities and by ensuring that those facilities are readily comprehended as existing within the interchange. Crime, too, thrives in overcrowded areas, in those that are poorly lit and in those that are not overlooked. There needs to be attention to the nature

4.23 Escalators, lifts and walkways lit by natural light make for safe and comfortable movement at Shanghai South Station designed by AREP. (Photo: AREP/T. Chapius)

of the spaces and the geometry of concourses and connecting routes to ensure that there is a safe feeling as perceived by vulnerable people.

Materials and finishes

The materials that are employed to construct the interchange should be robust and durable, attractive, appropriate to their purpose, easily maintained, and culturally and environmentally acceptable. Similarly, materials used in furnishings and fittings should share these characteristics but also provide comfort and sensory delight. Besides the normal criteria that architects employ, there are three special conditions that material selection needs to address at the interchange.

First, there is the obvious imperative of ensuring health and safety in the specification of materials. Slip hazard needs to be addressed, especially where surfaces may occasionally be wet,

as at building entrances. Special texturing of paving may be required, particularly at changes of level or in association with boarding transport vehicles. Besides the attention to able-bodied people, there are likely to be large numbers of elderly travellers, those with small children, and those with limited sight and personal mobility. There are a number of techniques employed to warn people of hazards, for example by stippling the surfaces underfoot or by adding studs that are easy to see and may glow at night. Being able to feel the effects in a tactile way provides reassurance, especially when tactile and visual qualities are combined.

Materials that are safe to use may not be so when they begin to wear. The fibres released may prove hazardous to health, either for passengers or those who are exposed for long periods, as are workers at the interchange. Rubber and certain synthetic floor finishes fit into this category. These and other materials may pose problems through the cleaning agents employed – the effect being a cocktail of hazards from a variety of cleaning chemicals used and physical degeneration activated

4.24 Smooth reflective surfaces are easy to clean and look smart and welcoming. New Zurich Metro. (Photo: Brian Edwards)

by the cleaning process itself. As a general rule, natural materials pose less of a problem than artificial and synthetic ones, and are generally easier to replace when damaged. As general rules also, materials and finishes that require little maintenance are those that wear well and look good for the longest period of time.

Second, people spend a lot of time at interchanges as they wait for their travel connections. Hence, they have time to study closely the finishes employed as they sit on concourse benches or wait in lounges. The quality of materials close at hand matters more than in most other design situations. The soft furnishings, handrails and paving matter particularly, as these are what is touched or closely examined. Hence, at Copenhagen Airport Interchange there is hardwood parquet flooring underfoot and furniture designed for style and comfort rather than utility. It creates a more favourable impression as a gateway to Denmark than branding by more traditional means achieves.

However, since transport interchanges attract high numbers of people, the finishes and furniture are subject to exceptional levels of wear. Typically, a European interchange such as St Pancras Eurostar Station in London handles around 23 million passengers a year and many other people visit the building to meet friends or simply watch the spectacle of travel. In the Pacific region, interchanges often experience even higher usage levels, with Shinjuku Station in Japan handling 2 million passengers a day or about 2 per cent of the country's total population. Since interchanges are also open to all and often accessible for 24 hours a day, they attract homeless people and others in social distress. Catering for their needs and related behaviourial problems adds a further level of difficulty for those who design and manage interchanges. The temptation is to reduce all design within vandalism radius to the bare minimum necessary. However, evidence suggests that such an approach simply invites more anti-social behaviour. It is better to follow the policy of high-quality initial design with ease of replacing damaged areas. This suggests an approach that recognises the need to invest in design quality for the message it sends regarding public transport (Grimshaw, 2009), but at the same time to plan for repair and upgrading worn-out or vandalised furniture, fabric and fittings.

Another consideration is to design construction details so that materials can be readily cleaned. Simple measures such as using coved skirtings, having glass areas that can be easily reached for cleaning, using hardwoods rather than softwoods, and carpets and linoleum rather than cheap synthetic substitutes, all enhance appearance and aid maintenance and cleaning. Toilets are an area that requires particular attention and they should be both robust and luxurious. Again, the use of curved junctions allows for more effective wiping by brush or cloth than traditional right-angled material joins.

The third factor to consider is how to express the flows of movement in the selection of finishes. Interchanges are places where interactions take place with the routes articulated or expressed in the choice of materials used in different parts of the building. The web of movement can be expressed in one material underfoot, while the spaces for waiting use different materials. Also, it may be possible to express the passenger movements to different travel modes in different materials, thereby aiding route legibility. Since there is much weaving of routes to aid inter-modality, the choice of finishes in the interchange can enhance the perception of the directional mobility. Some materials lend themselves to expressing better the weaving of the complex web of movement patterns in a typical large transport building. Ideally, the same material will be used on ramps and stairs as along malls. Moulded materials such as granolithic or terrazzo allow this to happen, possibly using different colour chips to identify the different routes involved.

Materials for travel should be different from those for sitting or shopping areas. Besides the wear element of surfaces subject to much foot traffic, there is the question of light reflectivity and of acoustics. Hard surfaces may be fine for foot travel but are tiring and produce a noisy environment within seating areas. Likewise, light-reflective surfaces are good for major routes but not for retail or more tranquil café zones, where subtle effects of light may be required commercially.

Materials and finishes design principles are:

- Use materials and finishes to help identify key movement flows through the interchange.
- Employ reflective surfaces on concourses, and non-reflective ones in waiting or commercial areas.
- Non-slip floor finishes are crucial, especially in wet/dry areas.
- Quality matters to the image of the interchange.
- Consider replaceability and salvageability.
- Consider ease of cleaning at design and specification stage.
- Consider the health risks related to worn finishes.
- Employ texture underfoot to guide those with limited vision.
- Source materials locally to provide cultural connections.

The same approach could be employed in the integration of indoor and outdoor spaces. Since the interchange is not necessarily a single structure, there is likely to be much interaction between buildings and outdoor spaces. There are also likely to be many changes of level and movement areas zoned for different needs. Here again the selection of finishes to paving, handrails and walls can help express the functional movement patterns and blur the distinction between inside and outside. After all, it is the patterns of people movement and flows of transport types that need to be expressed, not the abstract spaces in purely architectural terms. Materials can express the flow of people or traffic as effectively as signs and signal the different routes available along the access ribbons of a typical interchange. Hence, it is important that the architectural engineering, detailing and materials have a consistence in terms of design approach.

It is quite common to find a combination of heavyweight materials used in conjunction with high-tech lightweight ones. The former are required to satisfy engineering and acoustic demands, while the latter are employed for ease of replacement and the projection of a finely detailed appearance. Mass, frame and panel therefore make up the ingredients of the aesthetic language of many interchanges. Adding to this the demands of transparency and of lighting provides the opportunity to express materials and finishes in an architectonic fashion. As Grimshaw notes, his design practice (responsible for many of the major transport projects of the past two decades) has a 'strong ethic of concentrating on engineering, detailing and materials' (*Archidea*, 2009). Expressing the potential of

materials and doing so with an eye for proportion is important too, especially as transport buildings are subject to critical gaze while passengers wait for their trains, trams, buses, ferries and planes.

Another dimension to consider is that of consistency across transport networks. Normally, transport architecture is a linear affair with a number of facilities grouped along a route. There may be new interchanges to link together earlier provision, but often there are interchanges interspersed with a number of more singular stations or stops. The question emerges, then, as to whether the route needs visual coordination, using a common palette of materials and design approaches, or whether each station should have its own identity. Two recent examples can be used to illustrate the differences and their strengths. The first is the example of the new railway line linking central Oslo to Gardermoen Airport. Here, both the airport interchange and the linking stations of Lillestrøm, Eidsvoll and Asker use the same combination of slate, wood, concrete and steel, bathing the spaces in high levels of natural light while exploiting the potential for winter solar gain. Wood and slate

form the bulk of the finishes that catch the eye both inside and outside the stations. They are therefore identified by association as transport facilities, since there is a common set of materials and, hence, colours and shapes at each station. By way of contrast, the Jubilee line in London, constructed between 1995 and 2000, has a marked difference of architectural treatment at each underground station. Each architect for the separate stations employs individual colours, materials and effects of light. Although there is a consistency demanded by operational factors, the differences between each station are marked. Passengers are therefore able to identify a station by its architectural style and detailed treatment, unlike in Oslo where there is far greater uniformity. The Oslo stations have a corporate imagery, while the London example projects greater respect for architectural and cultural diversity.

Much current design practice sits somewhere between the two extremes described above. Architects today tend to design stations and interchanges so that they have their own image and identity, while acknowledging the value of those common elements that hold together the wider network. This helps

4.26 Light is important for animating surfaces, defining routes and expressing the quality of materials. Barnsley Bus Interchange. (Photo: Brian Edwards)

passengers understand the logic of movement and spatial organisation, and at the same time encompasses those distinctive elements that provide way-marking and site-marking. Such specific-ness can anchor the interchange into the mental map of passengers as well as the physical reality. Here, colour, effects of light and use of sculpture can provide points of orientation. If the interchange is above ground, opening up vistas to familiar landmarks in the wider streetscape can also create a sense of place as against placelessness.

Lighting and signage

Poor lighting and uncoordinated information systems are sources of frustration for many users of transport systems. Interchanges require particular attention to be paid to the display of information, to how areas are lit and to articulating the different movement zones through the building. Where changes of level occur, good lighting is critical to public safety and personal security.

Lighting at interchanges is by both natural and artificial means. Since many interchanges operate throughout 24 hours, the night-time lighting landscape is important. During the day natural light can be exploited to illuminate the bulk of the interchange. Both wall and roof lighting can be utilised, and both daylight and sunlight are available. However, whereas sunlight is useful in adding sparkle to the interchange and possibly highlighting major from minor routes, direct sunlight can cause glare as well as overheating. Older people are less able to deal with direct sunlight in their eyes than younger people, and glare at changes of level increases the risk of falls. Hence, sunlight penetration needs to be handled with care.

Another problem common to all transport buildings is that of moving between areas lit by natural means and artificial means. This occurs often at underpasses and at the transition between open platform areas and enclosed concourses. The

4.27 Well-lit volumes enhance architectural effects and improve the reading of direction signs. In this large glazed concourse at Frankfurt Airport Station, sunlight is carefully filtered. (Photo: Brian Edwards)

eye has to adjust to different light levels and often to subtle differences in the colour of light. The task for the designer is to maintain an even level of light and to ensure that the light effects during the day are similar to those at night. For example, information such as printed timetables may be perfectly legible by day but so poorly lit that they cannot be read at night. Likewise, concourse routes that are clear in daylight may be hard to find at night.

As a general rule, it is important that you can see into people's eyes (Juul-Sørensen, 2009b). This requires a relatively high level of illumination throughout – that is, without dark corners. One problem with lighting is that of maintenance: it should be possible to clean the lighting source whether natural or artificial and to replace exhausted light bulbs. Designing for maintenance is often overlooked, resulting either in neglect of essential cleaning or replacement, or the impeding of concourses while scaffolding is erected.

Lighting is a complex field with its own specialist literature. Architects need to utilise natural light wherever available, not just for energy efficiency but to provide for visual acuity and amenity. Sunlight, as mentioned, can animate concourse

spaces and add essential sparkle, but it needs to be handled with care especially around staircases. Solar heating of concourse spaces can also be exploited, with the benefit of solar cooling in the summer, but again the implications for route legibility and sunlight on display screens need to be considered. Also, since many large interchanges have extensive shopping and café areas, each with its own lighting regime, the problem of ambient heat from a mixture of artificial and natural sources needs also to be taken into account. In an age of global warming, the problem is likely to be excessively high temperatures, not unacceptably low ones.

Artificial lighting at night can add much relief to the drab environs of the typical interchange. Lighting can highlight the architectural structure, express atmospherically the nature of the different spaces, and illuminate in interesting ways passageways, malls or staircases. An example of the latter is the neon and laser beam lighting of the long walkway under the main runway at O'Hare Airport in Chicago. This passageway joins together two of the terminal buildings so it is heavily populated. Besides the theatrical effect of the different lighting colours projected on to walls and ceiling and hanging in the air, there is electronic music adding to the sense of fun. The example highlights the potential of artificial lighting to raise spirits and animate the transport experience.

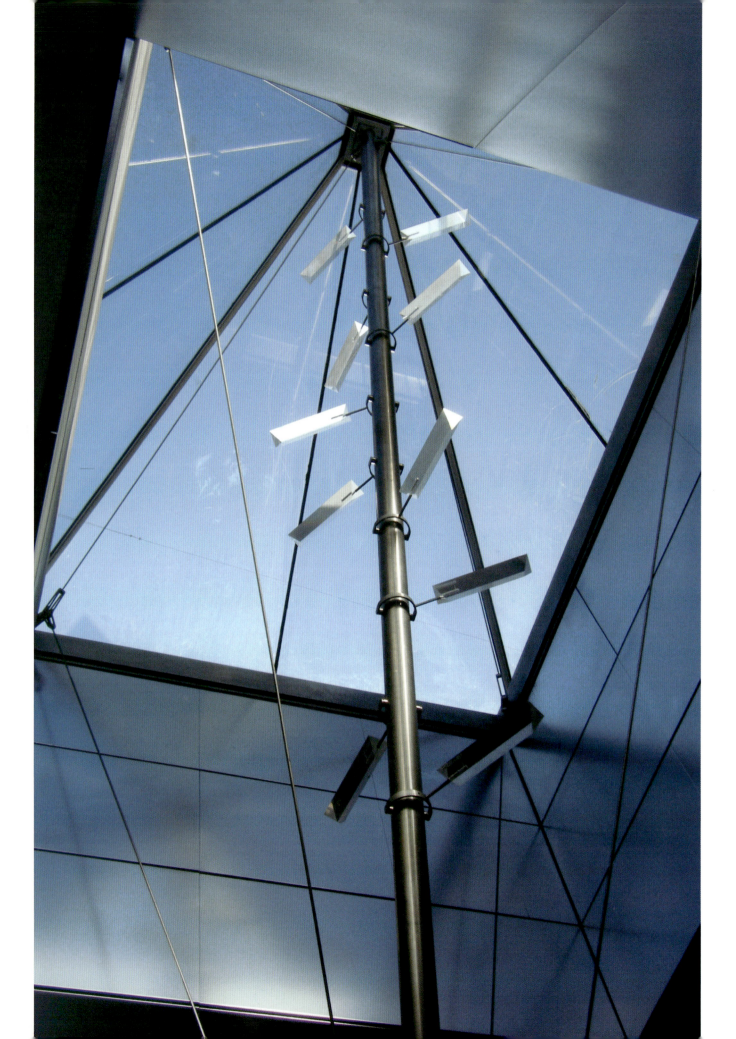

4.28 Rooflight with reflecting prisms at Kongens Nytorv metro station in Copenhagen. (Photo: Brian Edwards)

4.29 Art lighting at O'Hare Airport, Chicago. (Photo: Brian Edwards)

4.30 The Kyoto Station Interchange seeks to create a place rather than just space for travel. Expansive volumes help orientate passengers by opening up distant views of the city. Architect: Hara Koji. (Photo: Brian Edwards)

The outside of interchanges provides a further opportunity to exploit the potential of artificial light. Since the interchange is an important building in the wider cityscape, it needs to be illuminated in a way that draws attention. Here, designers can learn much from the retailing industry, where external flood-lighting, neon highlights and mixed colours allow one shop to stand out from another. Lighting at high levels also allows information screens to be read, helps with providing background security, and provides for safer use of the inevitable stairs, ramps and low steps that surround many interchanges.

Stairs present particular difficulty for the designer. They need to be rugged enough to withstand heavy use, yet elegant and pleasurable to navigate, and lit in a way that does not create shadows where one might otherwise expect to find steps. Lighting above the stair is commonplace, but this puts the steps

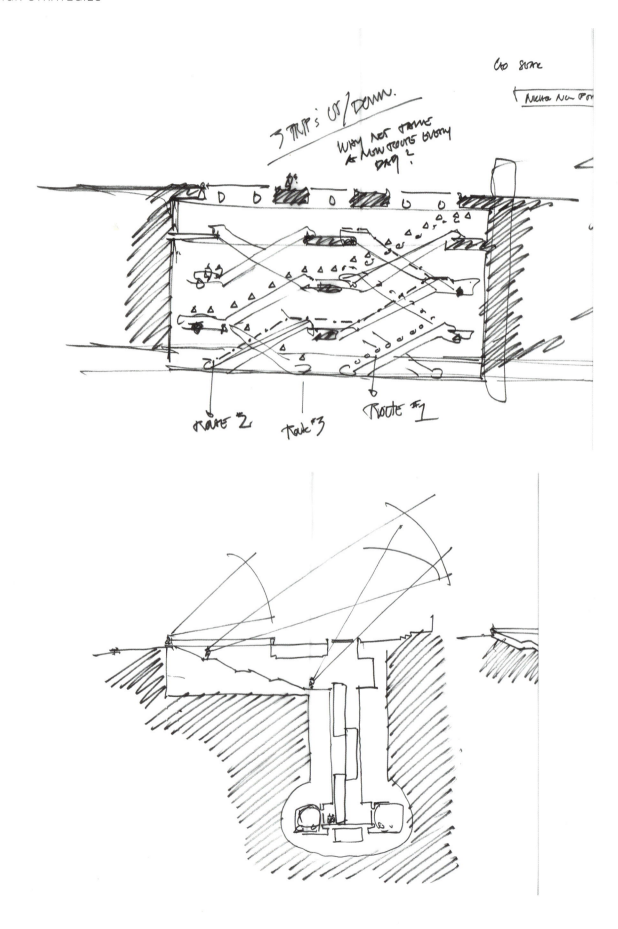

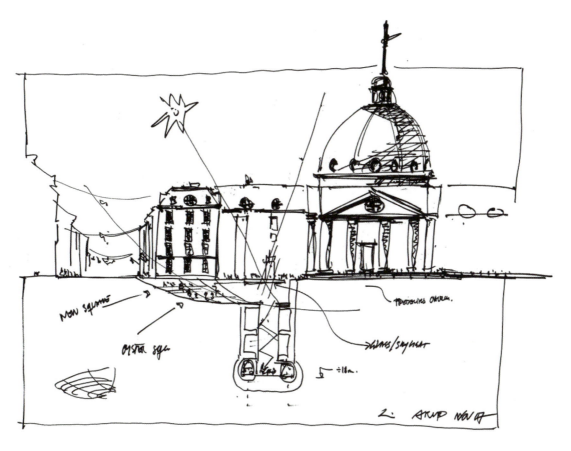

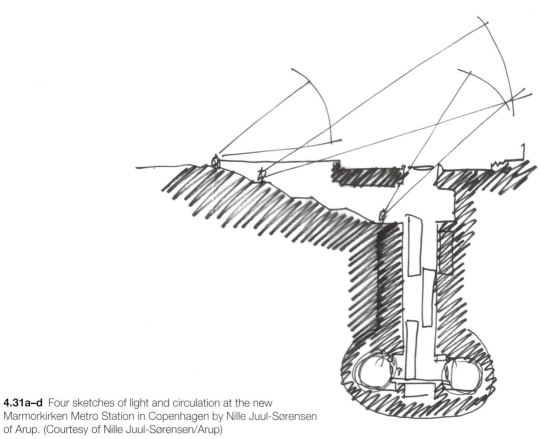

4.31a–d Four sketches of light and circulation at the new
Marmorkirken Metro Station in Copenhagen by Nille Juul-Sørensen
of Arup. (Courtesy of Nille Juul-Sørensen/Arup)

4.32 Art Installation at Schottenring Metro Station, Zurich.
(Photo: Brian Edwards)

into the user's shadow. Likewise, lighting from below (which is fashionable at present) means that, as one checks for stairs, the eye has to fight the uplighters to read the situation. Handrail lighting is also sometimes employed and has many advantages, but it rarely provides sufficient illumination on its own.

Since many routes through complex interchanges are diagonal, lighting and signs have to deal with a three-dimensional reality as against the simple movements that occur in transport facilities on only one level. Lighting, both natural and artificial, is useful in guiding people and reinforcing the message of directional signs. It is also advisable to provide a view of the desired destination so that people have a sense of achieving their goal. This may be merely a glimpse of a distant train or bus, or in the opposite direction a view of a familiar city landmark. Either way, a perception of the desired direction is more important than a handful of more abstract signs. Like big

airports, the interchange is a place where one is often lost in space, disorientated by shops and sounds, confronted by too many signs, and exhausted by the inherent confusion.

One trend is towards using glass panels in floors and pavements to illuminate areas below. This has the advantage of energy efficiency, since natural light is taken to areas otherwise dependent upon artificial light, and it provides a sense of day and night that is useful for long-distance travellers. However, caution is needed in areas where such panels may become wet, since they provide a slip hazard. It is not only the glass that is slippery when wet but the metal frame, which, being sometimes slightly raised, is both a trip and slip danger, especially for elderly people.

Since daylight is important in providing the perception of architectural spaces, the problem is one of designing for variable conditions. Daylight often involves sunlight, which is both a benefit (for potential solar gain and sense of well-being)

4.33 Clear signage and well-designed entrance at Rotterdam Blaak Station. (Photo: Brian Edwards)

and a problem (overheating and glare). Daylight is also of variable quality depending upon sky conditions and the time of day. Light is therefore of critical importance both above and below ground. It is, however, in the subterranean areas that particular attention needs to be paid to light. The challenge is how to bring daylight into areas many metres below ground. Here, there are three main strategies to employ. The size of the opening to the sky and the angles of the walls are important. Walls that are angled rather than vertical allow more light to be deflected into the underground platform and circulation areas. Having large openings at the top provides generous space for escalators, lifts and stairs, which can 'float' in the big volumes. The second strategy is to use mirrors, lenses and prisms to deflect and magnify lighting levels. These can be used functionally or artistically, but in either case the result can be more attractive and safer circulation and boarding areas. The third strategy is to consider the reflectivity of the materials employed, particularly on floors and walls. Unfortunately, the materials that reflect light also tend to be slippery and this is particularly a problem with polished floors. In many large transport interchanges all three lighting approaches are adopted, often

supplemented by imaginative use of artificial light as the spaces become more remote from natural conditions.

Another desirable principle to follow is to ensure compatibility between the design approach for natural light and that for artificial light. Although specific design solutions may hinder this, the perception of spaces and routes should share a common basis, whether during the day or night. Since space is the medium of connection at transport interchanges, how that space is lit affects legibility of key routes and hence directional orientation.

Good lighting within and around the interchange is essential for its operational efficiency. Passengers need to be able to see, to read the pattern of routes and spaces, and to identify their mode of transport. They also need to read signs (which are normally big), timetables (which are normally small), platform or gate numbers (which should be illuminated) and clocks. Knowing the exact time is important if journey connections are to be made. Sometimes clocks are only provided in analogue form and often in association with another system of travel information (such as timetables). It is better to have a full-face clock, centrally located inside and outside the interchange, illuminated and unobstructed by secondary advertising or

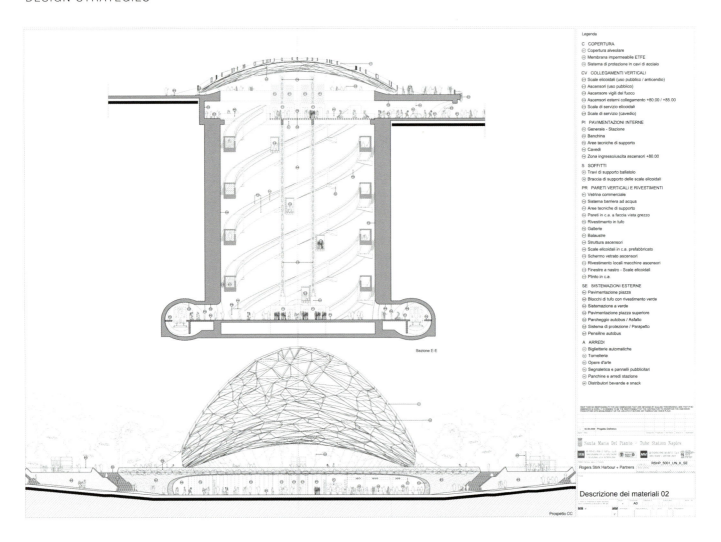

banners. Secondary clocks are also needed at each of the connecting transport systems.

A single information strategy, time-notification policy and standardised graphic display helps the traveller negotiate between different transport modes and their separate corporate philosophies. Although, in some respects, bus, train, metro and taxis are in competition for customers, they need to share in travel information packaging as well as in coordinating their different timetables. The strategies for space planning, time communication, timetabling, travel information and ticketing should share a common purpose at the interchange.

Baggage and storage

There are different traditions and different demands regarding baggage storage at transport buildings**.** Security and fears of terrorist bombing have also curtailed the opportunities for storage at most interchanges. However, to interconnect effectively between transport modes, storage of items such as skis, bicycles and wheelchairs is often needed. In many

Scandinavian countries it is common in the snow to ski to the local station in order to commute to work. Those that ski in the winter also tend to use their bicycles in the summer. Hence, there is a tradition of providing storage for personal transport modes at the station. More commonly, cars are parked at transport facilities, either casually or as part of park and ride facilities. The principle is the same: to transfer from a private travel mode to a public one, storing the former at the station for one's return.

How storage is provided varies, but normally there are three considerations: security from theft, safe transfer routes and facilities (including showers and changing rooms for cyclists) and good design. Car parks pose a particular difficulty in terms of landscape design, leading to some innovative solutions (e.g. Zaha Hadid's transport interchange at Strasbourg). Multi-storey parking is sometimes provided for cars, but more commonly for cycles, where elevated cages are often designed as part of the interchange. Good lighting and security cameras are required to reduce the risk of theft or vandalism.

Skis pose a particular problem. They are bulky and obstruct routes when casually stored. Many modern transport facilities

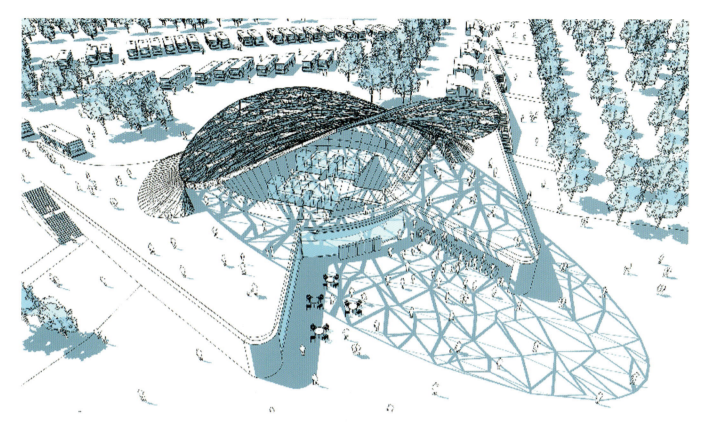

4.34a–b Section and elevation (a) of Santa Maria Del Pianto Station in Naples, designed by Rogers Stirk Harbour and Partners, and (b) CAD aerial view. (Courtesy of Rogers Stirk Harbour and Partners)

provide specially designed storage areas for skis and other bulky items. Whether it is cycle or car parking or storage of leisure equipment, it is important that the design brief includes recognition of the need to plan for such facilities. Too often, storage occurs as an afterthought, resulting in a poorly coordinated service for passengers and unsatisfactory design.

Involving artists

Art and architecture have frequently kept each other company in the transport buildings of the world. Good examples are in the Moscow Metro, where heroic sculpture stands inside and sometimes outside many of the grandly scaled stations of the 1930s, and in the Stockholm Metro of the post-war years. Since transport architecture is essentially a form of public space occupied by large volumes of people, it is natural that here one finds public art. Also, as waiting for trains or buses entails being stationary, art provides a means to entertain the passenger, raise deflated spirits and perhaps encourage a little internal reflection. A good recent example is at the Chatswood Transport Interchange in Sydney. Here, a 300 metre-long 'wave' wall in glass-reinforced concrete alludes to surfing and acts as a reminder of the ocean; there is also a work known as

'shareway' by sculptor Peter Cole placed in the public concourse and another work at the entrance to the bus station by artist Daniel Tobin. Their purpose is to aid navigation and help with the cultural interpretation of the public spaces (Chapman, 2009).

Art is frequently used to assist orientation. Large works, whether free-standing, suspended or in mural format, can provide points of focus for the traveller in busy concourses. This allows a mental map to emerge, which supports cognition and hence reassures passengers of the rightness (or wrongness) of the route they have chosen. Art in such situations needs to have a big impact and, as a consequence, most public art in interchanges is chosen for its ability to be memorable. In the Stockholm Metro, for instance, there is a unifying language of design with small variations at each station. However, it is the commissioned works of art that readily distinguish each stop on the line. Here, the art works have that blend of Nordic coolness and visual impact necessary to make a statement appropriate for each station context. Similarly, at the new Second Avenue Subway line in New York, designed within a common language of forms and components by architects Fox and Fowle, each station incorporates 'community-responsive art installations' (Wortman, 2005: 67). As with Stockholm and recent art installations in the new Zurich Metro, the aim is less about engaging passengers by confronting them with challenging modern art, and more about providing local identity through works that carry community acceptance.

PART 3

Examples from recent practice

The final part of the book shows how architects and engineers have applied the ideas in Parts 1 and 2 in practice. The selection of case studies is meant to represent contemporary design practice from different parts of the world. The emphasis is upon innovation at different scales, in different locations and in different types of interchange. Since the architecture of inter-modality is about moving between one form of public transport and another, the examples are necessarily hybrids of traditional bus, rail, ferry and airport terminals.

The cases are grouped according to the primary transport type. Hence, there are bus station interchanges where the bus is dominant, railway station interchanges where the train is dominant and so on. To be an interchange there must be at least two forms of public transport present, but their classification here is based upon the dominant mode. Hence, there are four chapters in this part of the book – on bus, train, ferry and airport interchanges. The bus interchange is placed first because it is architecturally the most rapidly evolving type and also serves a particular purpose socially. Historically, it is also the most neglected of the public transport modes, at least in terms of architectural scholarship, and also in terms of public financial support. After the bus interchange there are chapters on train, ferry and airport interchanges. Each has its own characteristics, which are illustrated via examples.

The aim is to describe the examples and draw some conclusions. One conclusion is the way in which national character finds its expression in the location, layout and details of construction found in many of the cases. For example, the high-speed train networks of Europe are leading to grand new stations in the French countryside (a kind of railway grand project), while in Germany the same investment is finding expression in remodelling existing major urban stations. Similarly, in China the new interchanges are an expression of a confident nation and one that is taking climate change seriously. In the USA, it is local action by city councils and state administrators that is leading the shift from car travel to metro, suburban and mainline train interchanges. Here, the initiative comes not from federal but local political action. In Japan and parts of Australia, attention is focused on ferry terminals and how they can better serve transferring passengers. Another pattern emerging is the way new transport buildings often incorporate new energy technologies in the form of photovoltaic panels and solar-assisted stack-effect ventilation. Public transport is leading to greener transport buildings, which in turn act as beacons of change within their neighbourhoods.

Many of the largest architectural and engineering practices are involved in the expanding market for design services in the arena of transportation. There are many familiar names in this section of the book – Arup, AREP, Grimshaw, Foster, Calatrava, FOA etc. There are also new and emerging practices, many of which bring interesting fresh approaches to the design of transport buildings. Innovation is a key quality sought by clients and by the travelling public, which has suffered from underinvestment for decades. Fortunately, the imperative of climate change and consequent reduction in fossil fuel use has redirected attention to public transport in general and to the interchange in particular.

Bus interchanges

Of the four main types of transport interchange explored in this book, the bus interchange is the most important socially and perhaps also the most difficult to design. Its social importance lies in the role of buses in providing mobility for the poorest members of society and of the primary role of bus transport in the cities of many developing countries in Africa, Asia and South America. Bus transport is what generally gels together the inner city – its web is pervasive and flexible and its costs are relatively low. Hence, the bus interchange contributes more to social sustainability and community cohesion than many other transport buildings. Poor people use buses to get provisions, to travel to work and to interface with other transport modes.

Buses are becoming more tram-like as they adapt to the needs of the twenty-first century. Many are thinner and longer (which helps with sharing road space with cyclists), have lowered floors (which helps with wheelchair, shopping basket and pram users) and run on electricity or green fuels. The hybridisation of bus and tram technologies opens up new possibilities for bus use in mature western cities, while widening the appeal of buses to middle-class users. There are implications for the design of road space, the bus stops themselves and interchanges used to transfer from bus to rail. New bus designs coupled with new information technology in the form of information, biometric ticketing and timetabling promise to change the shape of bus travel over the coming decades (Blow, 2005: 191–2).

In many ways, the bus interchange is the least glamorous of the interchanges discussed in this book. It often presents little more than a utilitarian face to the outside world, and often this is obscured by the many mainline buses and mini-buses that surround its outside walls. In large areas of the world the bus is unregulated and the bus station a place of competing services with little concern either for urban or building design, and as a consequence low priority is given to pedestrian movements. Sometimes bus stations are integrated into railways or urban metro systems, but generally in Africa, Asia and South America, the connection is on foot, or by rickshaw or bicycle.

In Europe the bus has a more important role within integrated public transport and often the bus is the first mode of transport before connection is made to other forms of public transport. The bus interchange is where the bus element is the dominant one within the transport hierarchy. As the principal mode, the bus will likely be at street level and highly visible within the interchange.

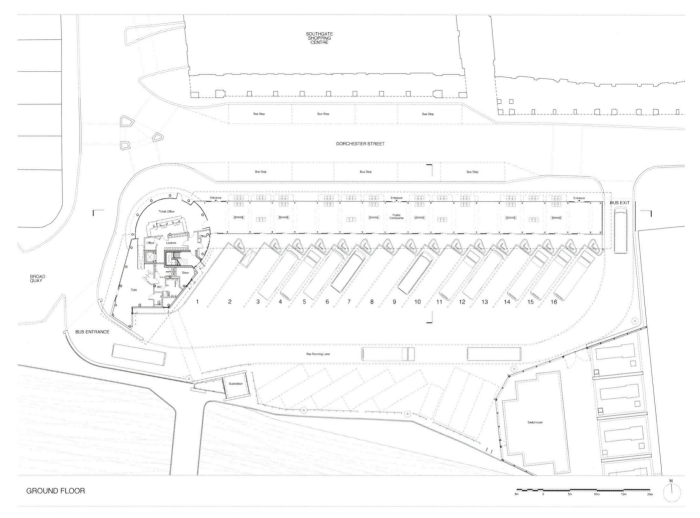

GROUND FLOOR

5.1a–b Plan (a) and cross-section (b) of Bath Bus Station, designed by Wilkinson Eyre. It has an elegance befitting this important eighteenth-century tourist destination. (Courtesy of Wilkinson Eyre Architects)

Other modes are usually above or below ground, or spatially a short horizontal distance away. Typically, urban bus interchanges interface with metro and rail systems. Often, with suburban bus interchanges there are large areas of surface car parking for park and ride facilities, but these examples are not discussed in any detail here.

As with all interchanges, it helps the passenger if there is a centralised system of providing travel information, and if this is provided in a number of locations both inside and outside the building. Good information reassures the passenger and provides breathing space for buying a newspaper or a cup of coffee. Ideally, too, the different travel companies will accept each other's tickets so that people can choose the mode of travel between, say, a bus, metro or local train service. Single ticketing, integrated timetabling and coordinated travel information at the interchange reduces the stress of travel. No matter how well the travel vehicles or travel buildings are designed, if there is tension caused by poor information or

unnecessary queues at ticket barriers (due to lack of recognition of the tickets of different travel providers), public transport will not be able to compete with private cars. Nowhere is this more important than at the bus interchange, where movement cycles are fast and information most fragmented between transport types.

Buses suffer more than most other forms of public transport from congestion. As a result, the bus interchange runs less smoothly than other types and generally has a more chaotic air. Bus interchanges operate more effectively when there are dedicated bus lanes nearby, but these have the effect of restricting access by cycle or taxi. Pedestrian movement can also be problematic, since buses operate at street level rather than high or low level as with metro and rail systems. People movement and bus mobility can be in conflict, especially immediately around the interchange where reversing may be taking place. Buses also are a source of air pollution and noise, adding to the perception that bus travel is at the bottom of public choice.

Bus interchanges have the task of resurrecting the reputation of this beleaguered transport mode. Design is central to

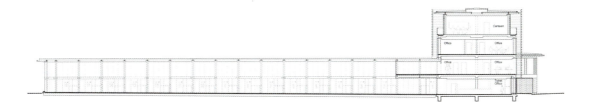

LONG SECTION

meeting social need and in addressing poor image. To attract the maximum number of passengers, buses should inter-connect with rail, metro, cycle ways and pedestrianised streets. Although not many cyclists transfer to buses (since many bus journeys are within bicycle travel range), the bus interchange should provide storage space for cycles. All bus journeys start and end on foot and, while there may be transfers, the pedestrian must be given safe, attractive, well-sheltered and barrier-free movement within and around the bus interchange.

Another characteristic of the bus interchange is the lack of retail space found in other types of transport interchange. Since bus journeys are normally of short distance and limited time, passengers do not need to buy food, newspapers or drinks to the same degree as those using other modes. Hence, bus interchanges lack the kind of commercial areas found in transport buildings more dependent upon dwell time and big journeys. Also, since arrival by bus lacks the social cachet of plane or high-speed international train, there is less romance. Bus interchanges are highly functional places where to change transport mode quickly is the primary objective. However, in pursuit of diversifying income, many recent intercity bus interchanges have introduced retail areas, borrowing the basic geometry from the shopping mall. One benefit of this is the potential of the mall to direct passengers towards the different transport modes present. When a variety of transportation types exist at the interchange, the mall needs to take a non-linear form in order to avoid the shops obscuring passenger routes. In these situations the mall should be cruciform in shape or set to one side of the main passenger concourse. Another useful device is to ensure that malls are lofty where passenger circulation occurs and less high in the shop areas – this symbolism, common to high street retail malls, allows light and ventilation to be accessed at high level while also signalling the major routes.

North Strasbourg Bus Station and Interchange

Designed by Zaha Hadid, the Hoenheim Bus Station and Interchange in North Strasbourg won the European Union Prize for Contemporary Architecture (Mies van der Rohe Award) in 2003. The interchange is mainly a large park and ride facility with associated links to bus, tram and railway systems. The interchange expresses the transition between transport types as a series of overlapping planes, lines and travel spaces. Hadid has explored, and to a degree reinvented, the architecture of transportation, using the familiar repertoire of multi-faceted shapes and fluid spaces. Fluidity is a key quality of transport buildings, with their emphasis on mobility, connection and social engagement. This building expresses the very nature of interchange not so much as function but as a performance space for users and as a spectacle for observers. Besides architecture, there is public art and lighting and landscape dealt with artistically.

The programme is relatively mundane and the site typically nondescript. The interchange sits at the end of one metro line, is crossed by another and has a future rail link planned. Buses occupy a major part of the present interchange, with trams located more towards the perimeter. Passengers wait in a central concourse sheltered by a dramatic angled, cantilevered roof. To one side are buses, to another the terminating tramlines and, at right angles towards the southern edge of the structure, is the train line. The main spaces are open to the elements but sheltered by the sloping concrete roof and angled walls. The trapezoid roof has large cut-outs to let sun into the concourse below and at night these are lit by clever effects of coloured light.

The plan figure is distorted by lines that draw their inspiration from the approaching angles of the trams and buses and from

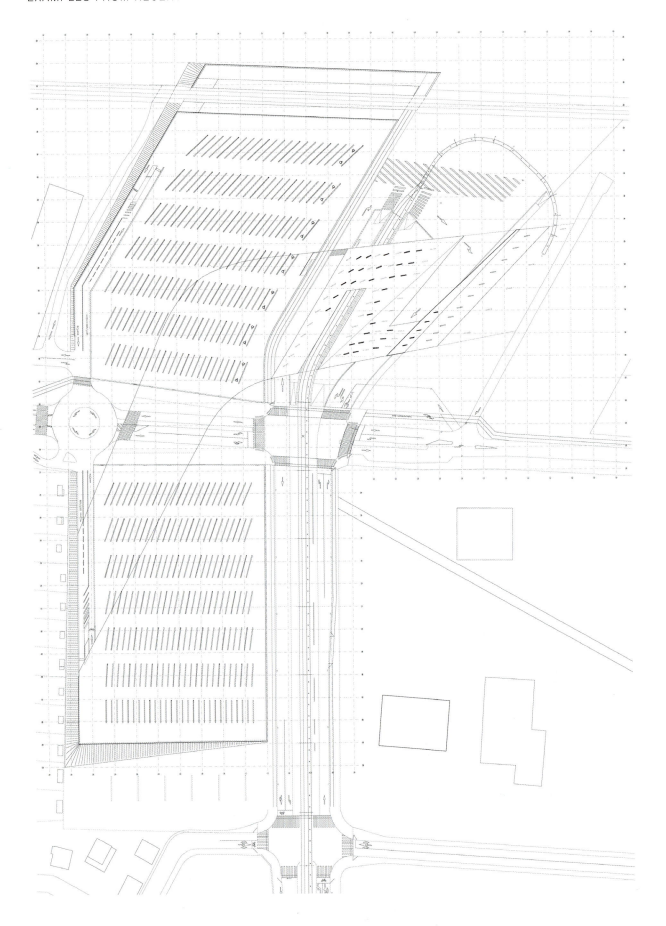

5.2 North Strasbourg Bus Interchange, designed by Zaha Hadid. (Courtesy of Zaha Hadid Architects)

the wider landscape. The waiting areas sit between non-orthogonal forces, creating a tension that is further expressed in the angled roof with its large cut-outs, freely arranged strip lights and steeply inclined columns. The sitting areas are reminiscent of the passage of mechanical transportation, with the play of stark contrasts between light and shade, static and mobile, landscape and the city. As such, Hadid's creation breaks new ground for the type, just as Foreign Office Architects reinvented the ferry terminal in Yokohama.

This is a relatively modest interchange with a suburban rather than city scale. However, the treatment of the ground plan and of the cross-section elevates the building into something of a local landmark. This is partly expressed in the exaggerated cantilever of the roof canopy, but mostly in the clever interplays of landscape and architecture. The latter derive their justification from the very concept of interchange as space for social as well as transport interaction. Although the materials, mainly concrete with steel columns, are harsh and unforgiving, they lend themselves to robust and sculptural interchange architecture.

Vauxhall Cross Bus Station and Interchange

The Vauxhall Cross Bus Station, designed by Arup Associates, is linked to Vauxhall railway station and underground just south of central London to form a new transport interchange. The project, which cost £4.5 million in 2005, is an attempt to make a coherent and efficient connection between various forms of public transport in this busy corner of London. Unfortunately, the scale of road provision around the interchange and associated volumes of vehicles using the perimeter road system limits the effectiveness of the interchange as a hub.

Vauxhall Cross is designed primarily as a bus interchange with links to the elevated rail and underground systems. Buses converge and circulate here using a perimeter road system, with passengers in a linear centre 200 metres long of dedicated bus stopping bays and associated seats and information boards. They are protected by an undulating steel canopy 12 metres wide, which acts as an emblem for the project. At its south end, the canopy splits into two inclined fingers, which house photovoltaic panels intended to provide electricity for

lighting the interchange. The PV cladding complies with the Mayor of London's energy strategy (which requires all public buildings in the capital to generate 10 per cent of their electricity from renewable energy) and here was part funded under a central government PV Demonstration Programme. The 200 square metres of PV panels provide 30 kWph over a typical year (Fraser et al., 2005: 3–6).

Built to deal with 2,000 bus movements per day, the interchange is now the second busiest bus station in London, with over 50,000 passengers using the facility every day. Usage has increased by 40 per cent since the interchange opened in 2005, indicating the importance of public transport interconnection to quality of life. The design was undertaken in collaboration with police crime prevention experts and contains good sight lines for surveillance by transport staff, high levels of transparency and good night-time lighting (Jones, 2006: 66). As a result of the success of the Vauxhall Cross Interchange, adjoining areas are being upgraded faster than anticipated and plans for improving other transport facilities in the area are being brought forward.

The stainless steel undulating canopy is artistically lit in order to act as a landmark for the facility. The dips in the canopy undulations mark the areas where seating occurs and the high points signal where the double-decker buses stand. The canopy makes only limited contact with the ground in order to avoid obstructing movement, to prevent places for muggers to hide, and to aid legibility for the many elderly users of the interchange. Structurally, the canopy is made of two parallel 'I' beams made of repeating modules that were clad and painted prior to erection. Public consultation prior to design highlighted the need for shelter, safety, retail and toilet facilities as well as colour and attractive finishes. The result is a well-considered fusion of engineering, public art and transport facilities that is helping to regenerate this corner of London.

Barnsley Interchange

Constructed as part of the Remaking Barnsley Strategic Development Framework, the bus and rail interchange acts as a landmark of regeneration in this former Yorkshire coal-mining town. Designed by Jefferson Sheard Architects, the aim is

5.3 Vauxhall Cross Bus Interchange, London, designed by Arup Associates. (Photo: Brian Edwards)

5.4 Waiting area beneath canopy at Vauxhall Cross Bus Interchange, London. (Photo: Brian Edwards)

5.5 Barnsley Interchange, Yorkshire, designed by Jefferson Sheard. (Photo: Brian Edwards)

to create a modern transport environment that will act also as a meeting place with associated retail and office units. The solution adopted is based upon a realigned road system with a high-level bridge taking passengers between the bus and train station. The bus element is large and has a central curving mall with shop units extending from the glazed core to the external street. Bus stands face on to the mall on the opposite side from the shops, creating a well-functioning and lively pedestrian environment.

The architects sought a mixed-use interchange and drew funding from various sources to pay for the diversity of facilities available. A private developer provided funding for the retail and office components, the Barnsley Metropolitan Borough Council, using European Union funds, provided a great deal of infrastructure capital, and the bus and rail operator, in the form of the South Yorkshire Passenger Transport Executive, made its contribution. The result is a successful example of public transport-led urban regeneration.

Pedestrian routes and entrances are particularly well signposted, using space and transparency as the medium of connection. Inviting glazed gateways draw passengers and shoppers into the interchange and their routes are then signalled by high, glazed streets. These are roofed in EFTE rather than glass because of its light-transmission qualities, recyclability, toughness, lightness and safety record. In the event of a fire the material melts safely. Since EFTE is considerably lighter than glass, the structural elements are reduced in size and cost. Here, the roof is supported by timber columns and trusses supplied by a reliable eco company.

5.6 Concourse, Barnsley Interchange. (Photo: Brian Edwards)

Clad extensively in copper (chosen for its durability, ability to meet strict radii criteria and reuse potential), the interchange forms an elegant landmark in this north English town. Positioned as a pivotal building within the masterplan for regeneration in Barnsley, the contemporary look of the interchange provides an attractive gateway building to visitors. At a cost of £24 million, the bus element of the interchange draws many people to a part of the town that was previously a largely run-down industrial area alongside the railway station. One key element in the success of the project is the way the shop units provide a welcoming outer edge to the bus station. The disadvantage of hiding the bus facilities by such an arrangement has been overcome by the positioning and size of the entrances to the internal mall and the connecting bridge to the station. The latter signals the secondary entrance with considerable panache.

Manchester Transport Interchange

Designed by Ian Simpson Architects in collaboration with Jefferson Sheard Architects, the Manchester Transport Interchange connects together bus, tram and park and ride facilities. The arrangement is unusual – a two-storey interchange sits beneath a multi-storey car park, the whole composition sitting on a triangular island site on the south side of Manchester's city centre. Besides interchange and ticketing services, there are retail units and a café within the centre of the development.

From the outside, the impression is one of concrete ramps and brick and tile walls topped by a green glass-clad car park. The interchange has some of the style associated with north of England brutalism, although here it is tempered by careful selection of materials and colours. The effect is rather more sculptural than architectural, especially when seen within the wider cityscape. Entrances for pedestrians are marked by clever breaks in the perimeter wall and a long curved canopy that unifies the tram–bus interface. People are encouraged to park above and transfer to bus or tram services without leaving the building. It is a clever concept, but it results in transfer facilities that are located in the centre of the building. Hence, the perimeter is given over to circulating buses and trams, and via the ramps also to circulating cars.

Inside the interchange there are two islands for bus connection. The larger one contains separate areas of retail and café units plus a control room and travel shop. The plan is rational and, since the passenger accommodation has irregular shapes and rounded corners, the interior is not without architectural interest. Between the islands views have been created to the transport modes and to the city beyond. The tram (known as Metrolink) runs on the south side of the interchange and has its own island for passenger connection. It is joined to the bus element by a glazed, steel and copper-clad canopy shaped like a boomerang.

Mölndal Interchange, Gothenburg, Sweden

The Mölndal Interchange is an interesting mixture of high-level bus station and low-level suburban rail. It functions as a commuter interchange in the suburbs of Gothenburg to designs by Wingårdh Architects yet, in spite of its mundane function, the interchange acts as an important local landmark (Jones, 2006: 118). The interchange straddles an elevated highway, which brings buses to a high-level deck placed roughly at right angles to the suburban railway. The deck is cut away in order to allow light to reach the waiting passengers. The bus station has a café placed against a large sloping wall, which helps signal the location of the interchange, not unlike the use of the angled roof at Vauxhall Cross.

The bus station acts as a bridge between rail and bus movements and, since it straddles the railway tracks, provides elevated access to the train platforms. It is connected to timber-clad canopies that shelter bus passengers and provide the waiting bays for different bus services. The overall impression is one of civic pride in the provision of public transport. The use of clear and coloured glass, metal cladding and timber slats, supported by concrete columns with a secondary steel frame, breaks with the brick monotony of the neighbourhood. The composition of sharply angled shapes is suggestive of the work of the sculptor Richard Sierra and, being elevated, the interchange stands as a beacon, particularly at night.

Table 5.1 Bus interchanges

Name	Type	Architect	Key features
Vauxhall Cross Bus Station and Interchange, London	Bus/metro/rail	Arup	• Sculptural form • Civic space
Central Bus Station, Munich	Bus/rail	Auer + Weber	• Distinctive curved shape • Civic square
Mölndal Interchange, Gothenburg	Bus/tram/train	Wingårdh	• Distinctive shape • Wind barriers
Box Hill Bus Station, Melbourne	Bus/train/tram	McGauran Soon	• Three-storey interchange • Integrated with existing retail mall
North Strasbourg Bus Station and Interchange	Bus/metro/rail/ park and ride	Zaha Hadid	• Zigzag plan shape • Big-angled cantilevers • Won Mies van der Rohe Award
Barnsley Bus Interchange	Bus/rail	Jefferson Sheard	• High levels of daylight and ventilation • Recycled materials • EFTE roof through retail units
Manchester Transport Interchange	Bus/tram/ park and ride	Ian Simpson with Jefferson Sheard	• Sculptural form • Clad in green glass • Island site

5.7 Mölndal Interchange, Gothenburg, Sweden, designed by Wingårdh Architects. (Photo: courtesy of Ulf Celander)

5.8 Site plan, Mölndal Interchange, Gothenburg, Sweden, designed by Wingårdh Architects. (Courtesy of Wingårdh Architects)

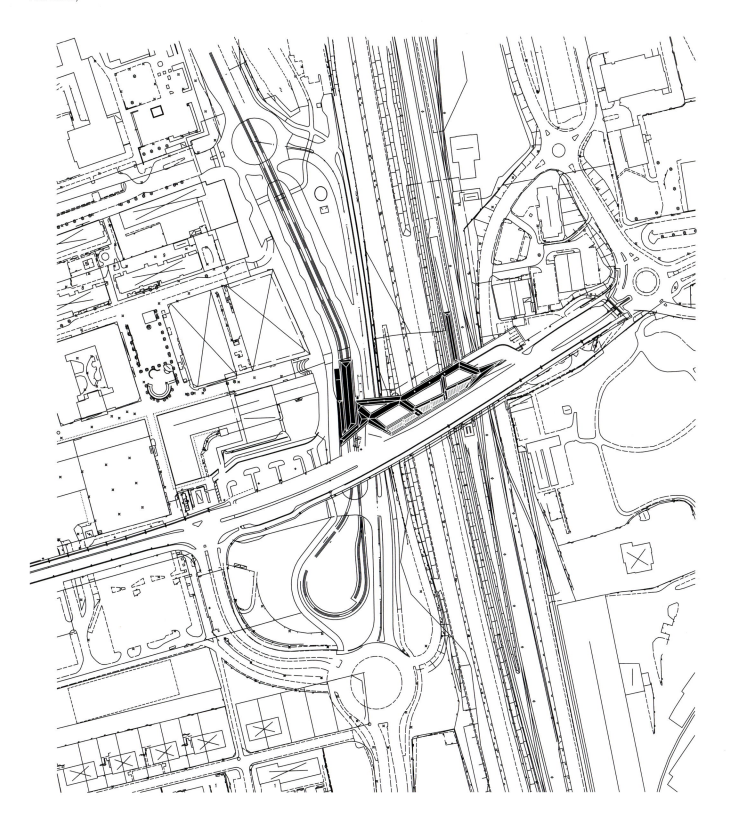

Train interchanges

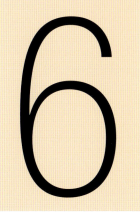

CHAPTER

6

Train interchanges are one of the most common and complex types of modern interchange. Since major stations are usually found in city centres, they are increasingly seen as inter-modal connectors. They often involve the extension to existing structures, many of which are listed buildings or are found in urban conservation areas, resulting in much attention to contextual design. They also frequently entail the jigsawing together of old transport infrastructure and new, often smart, movement technologies such as light rail and guided buses.

The growth of high-speed trains also stresses existing stations. Such trains are normally longer than traditional trains (sometimes 400 metres long), their passenger numbers increase the demand for transfer to other transport systems, and their popularity for mid-distance travel means that many more business journeys are conducted by rail than in the past. Hence, existing stations, particularly those in the bigger cities, are currently undergoing transformation to enable them to fulfil their new role as interchanges. Extending stations, as at St Pancras in London and Dresden in Germany, allows for wider urban reconstruction and this encourages the regeneration of often neglected urban areas. Hence, social and economic sustainability are often associated with new rail investment.

Underground railway interchanges pose four particular difficulties: how they are going to be lit and ventilated; how way-marking in cavernous spaces is to be achieved; how connection to the surface systems is to be achieved; and how the interchange is to be identified within the urban scene. The latter is important, since the life of the street rarely acknowledges an underground facility of any description, let alone a subterranean transport hub. The engineering of deep-cut transport facilities imposes its own discipline upon surface design. Cut and fill has major impacts on pedestrian movement, disturbance to road traffic and retail success. Tunnel boring is less environmentally damaging but usually entails deeper tunnels, resulting in greater vertical travel distance for users and less opportunity for taking natural light and ventilation into the interchange. With many interchanges, part of the system will be at ground level and part under the surface and, hence, visual connection becomes an important design consideration. Where the major hub is below ground, the architect and engineer will need to work closely together to meet an acceptable solution for users as well as transport providers. Interior designers, in particular, will not thank those infrastructure engineers who put the needs

6.1 Swansea Interchange, designed by BDP. (Courtesy of BDP)

of track and train above human needs. However, underground interchanges open up the opportunity to shape space out of rock in the manner of a sculptor exploiting the materiality of the rock face.

Vertical movement is usually by mechanical means first and by foot second. Hence, lifts and escalators will need to be provided as well as generous staircases. Since they take a lot of space, they cannot be accommodated within the pavement zone of existing streets. The usual answer is to create a square or piazza at the interchange ground entrance with the vertical means of access dropping down from this space in a functionally and visually coherent fashion. Normally, the ticketing and bridging zone is 6–7 metres above the track, which results in a mezzanine format midway between ground and track level. Here, the public facilities of shops, cafés, toilets, travel information and ticketing are located with a roughly equal distance vertically to the street above and the tracks and platforms below. So the role of the urban square is to signal the location of the routes down to the interchange and to handle the surface movements both on foot and by wheel. Often a

structure will be required to house the stairs and lift and this will need to be eye-catching in order to aid use legibility.

The following examples are typical of the emerging architecture of the station interchange. Two trends are apparent. The first concerns the different outlooks in different regions of the world. In China, the trend is towards large and monumental transport hubs where the needs of infrastructure are dominant. As a result, people needs are somewhat secondary, particularly in the area of pedestrian flows and cycle storage. In continental Europe, on the other hand, the social networks are generally quite strong and this often leads to infrastructure compromise. People here can move smoothly from urban spaces to interior transport ones, but sometimes finding a bus or metro connection can be difficult. As a general rule, there is strong civic leadership in transport provision in Germany, Spain and France, which leads to speedier implementation than in the UK with its cumbersome planning system and strong heritage lobby.

The second trend is one of reconnection. In Britain and the USA, investment has focused on better connection of existing

6.2 Aberystwyth Interchange, designed by BDP. Notice the attention paid to pedestrian routes. (Courtesy of BDP)

services through the provision of new or upgraded transport links. Here, new transport structures have been built to make existing buildings, facilities and transport systems more effective (e.g. St Pancras/King's Cross Interchange in London). Designs such as for Fulton Street Transit Center in New York and Stratford in London (gateway to the Olympics) seek to provide a sense of arrival as well as departure. Too often in the UK, examples of interchanges are hampered by the compromises made in the past, resulting in facilities that compare unfavourably with mainland Europe and the emerging economies of the Pacific Rim.

St Pancras/King's Cross Interchange, London

The decision to bring the high-speed train services from France into the UK via the Channel Tunnel to the Stratford connection in East London meant that St Pancras Station was the natural place to form an interchange between international high-speed rail, national and regional rail services, and underground and bus services. The interchange is a complex connection of many existing transport buildings, plus major additions and substantial urban re-ordering. Since many of the buildings are listed because of their historic and architectural interest, the room to carry out drastic changes was limited. However, the result is an interchange that joins together the complex geometries of existing transport provision at ground level and beneath, but much remains to be done to calm traffic adjacent to the station

exits and to remove the many informal structures that have grown up in the public concourses.

The masterplan for the redevelopment of St Pancras Station was drawn up by Foster + Partners in 1999, with John McAslan + Partners providing design frameworks for the adjacent station of King's Cross. Both gave emphasis to clearing away clutter around the two major historic railway stations in order to open up their façades, to provide better surface connection, and to clarify routes for pedestrians. Linked to this was a strategy of better connecting the underground to above-ground railway and bus services. Space and route legibility underpinned both masterplans, which had the difficult task of coordinating a multitude of different forms of investment while phasing the construction in order to allow for the continuing use of the interchange by 66 million passengers a year. The main development was undertaken by a consortium of Bechtel, Arup, Systra and Halcrow, employing a number of different architectural practices including John McAslan + Partners (at King's Cross), Allies and Morrison (underground links), and Pascall + Watson (interior design and restoration of historic buildings).

Being both termini and interchange, the St Pancras/King's Cross railway stations do not have the disadvantage of their sites being bisected by railway lines. However, a tradition of using the station forecourts for through road traffic has resulted in a lack of connection between the new interchange and the city of London to the south. The pedestrian has to negotiate a series of tunnels or wait at traffic lights to cross the busy Euston and Marylebone Roads. In this sense the St Pancras/King's Cross Interchange

6.3 View of King's Cross redevelopment by John McAslan + Partners. St Pancras is to the far right. (Courtesy of John McAslan + Partners)

6.4 View of forecourt at King's Cross Station as redesigned by John McAslan + Partners. (Courtesy of John McAslan + Partners)

6.5 St Pancras redevelopment model by Foster + Partners. (Courtesy of Richard Davies)

compares poorly with similar projects in Germany, such as at Leipzig and Stuttgart with their major external civic spaces.

Besides their handsome façades, the two main railway concourses are noted for the scale and grandeur of their interior spaces. At St Pancras the iron and glass train shed designed in 1866 by the engineer Sir Thomas Barlow forms the main thoroughfare for pedestrian movement to the high-speed and domestic rail services. An elevated concourse overlooking the Eurostar services provides a café, shopping and viewing area. Since the high-speed trains have a length of 400 metres, the original station platforms have been extended, using a design that departs from the original arrangement. This distinction helps signal the location of the domestic, as against international, services.

As St Pancras Station was built above ground level in order to cross the Regent's Canal, there is an extensive undercroft. This originally served a variety of warehouse and office functions, but has been converted at the interchange into public concourse areas that link across the station at low level. Lined in brick and arched, these add considerable character to the development.

As with many interchanges, there are a number of entrances and connections at different levels. Since St Pancras and King's Cross adjoin at different angles, many of the public spaces lack orthogonal shape. This adds to the complexity of the whole and has encouraged the use of a very limited palette of materials, mainly brick and natural stone, in order to unify old and new, and those spaces that are regular to those more fragmentary in shape. The triangular space between the two main station side walls is occupied by a hotel, which adds further to the difficulty of unifying the whole into a coherent interchange. However, the main connections between the underground railway station at King's Cross (the subject of a major fire in 1987 and not far from the terrorist bomb attack of 2005) have

6.6 High-level walkway through original platform concourse at St Pancras Station. (Photo: Brian Edwards)

6.7 New concourse and platforms at St Pancras Station. (Photo: Brian Edwards)

6.8 New western entrance to St Pancras Interchange. (Photo: Brian Edwards)

been much improved, partly due to opening up the floors of the existing station in order to let natural light filter down to the concourse spaces below. Also, the 1970s' front extensions to the mainline King's Cross Station are planned to be removed, creating a piazza here for meeting and promenade. Foster's emphasis in the masterplan on clarity of movement and space legibility has encouraged such actions by subsequent architects.

At peak times the interchange caters for 82,000 passengers per hour. This considerable volume of movements, between 5 main railway services, 6 underground ones and about 12 bus lines, is a complex undertaking logistically. It involves large areas for ticket offices, for queuing and waiting, and for shopping. The plan of the interchange suggests that, of the total area, about 80 per cent is internal spaces of various types with circulation accounting for nearly 60 per cent of this. Thus, circulation receives more space than at other interchanges, reflecting both the volume of people movements and the complexity of the interconnections.

Stratford Interchange, London

Masterplanned by Foster + Partners, the Stratford Interchange is currently being expanded to cater for the London Olympics of 2012. The work consists of expanding the existing provision of connecting mainline, underground rail and bus services into a larger complex that will act as a gateway to the Olympic site. Besides the renewal of the area, prompted by the decision to locate the London Olympics nearby, two earlier infrastructure developments necessitated the creation of a large interchange at Stratford. First, the decision to create an international station here on the Eurostar route between the Channel Tunnel and St Pancras gave the opportunity to unify the fragmented transport provision in the area. The second was the more recent decision to extend the Jubilee underground line northwards to Stratford. So, by 2012, the new station has to serve as an interchange between internal and domestic travel, between rail and underground, and between rail and bus services, and also be a welcoming gateway to the Olympics.

The overall design by Design Rail Link Engineering, a consortium of infrastructure providers, engineers and architects,

is based upon the creation of a new and refurbished square surrounded by transport services, shops, restaurants and public facilities. Other architects involved include Wilkinson Eyre for the station element and London Transport's own design team for the bus terminal. The new urban square is crossed at high level by a station concourse where tickets and information can be obtained, and at mid level by the high-speed (the outer edge) and domestic (in the centre) rail services, and is served by a square to one side of the bus station, beneath which run the underground metro services. The station, designed as a grand airy bridge by Wilkinson Eyre, is highly glazed, providing visibility of the transport services beneath and views across this area of London to the new Olympic facilities. Passengers move vertically through the interchange via a central atrium, which also delivers natural light and ventilation into the core of the complex. Three principles have been followed in the design: to provide visual linkage between the different services; to maximise natural environmental conditions in order to save energy, especially through the use of daylight and sunlight to exploit stack-effect ventila-

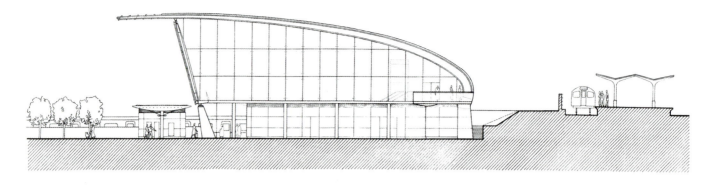

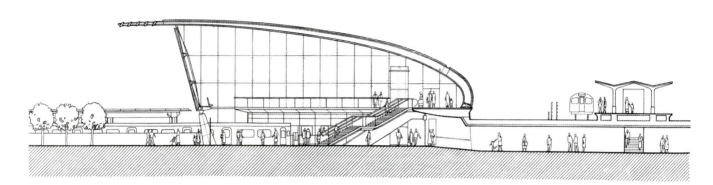

6.9 Section of Stratford Station Interchange, designed by Wilkinson Eyre Architects. (Courtesy of Wilkinson Eyre Architects)

tion; and to design the interchange as a civic focus rather than just a transport hub. One area of particular interest is the use of inverted cone-shaped canopies to shelter passengers waiting at the bus station. These collect rainwater and are lit from beneath, producing a distinctive lofty canopy for the alighting double-decker buses.

The new interchange is designed to handle nearly three million passengers a year, many of whom will be international travellers using Stratford as their point of arrival in the UK. Therefore, the facilities are engineered to project a good image of Britain's architectural and industrial design, using the familiar language of transparency and high-tech. The architects have sought a quietly elegant solution that is welcoming by both day and night, and one that uses a range of repetitive and vandal-resistant materials and details in order to keep future maintenance costs down. As an international gateway, there are customs, immigration and security areas located beside the Eurostar departure lounge.

Birmingham New Street Station

The proposed refurbishment and extension of Birmingham New Street Station by Foreign Office Architects (FOA) is typical of attention being paid to the unsatisfactory state of many of Britain's older stations. New Street Station was built in the

1960s to accommodate about 60,000 passengers a day, while today the same infrastructure caters for 140,000, many of whom use the station as an interchange. The redevelopment provides space and facilities for 52 million passengers a year, with improved connection to other transport services. Such is the stress on station facilities that, in 2009, platforms had to be closed on 20 occasions due to overcrowding, with the concourse nearly coming to a standstill.

The striking design was developed by FOA working as part of an integrated project team, including Atkins and Mace, under the leadership of Network Rail – the UK's provider of railway infrastructure. The project is known as Birmingham Gateway, a term that signals the importance of the station as a point of entry into Britain's second biggest city. The total cost of the build is £370 million with funding provided by Birmingham City Council (via the Department of Transport), Network Rail, the British government-funded regeneration agency Advantage West Midlands, and Centro. Part of the justification for the inclusion of such a wide portfolio of funding bodies was to ensure that the new station acted as a focus for wider environmental improvements, including better transport connections and the regeneration of the areas surrounding the station.

Under the plan, the passenger concourse will be tripled in size, an atrium will bring natural light into the passenger spaces, and new lifts, escalators and stairs will be provided. There will also be several new entrances created in an attempt to improve

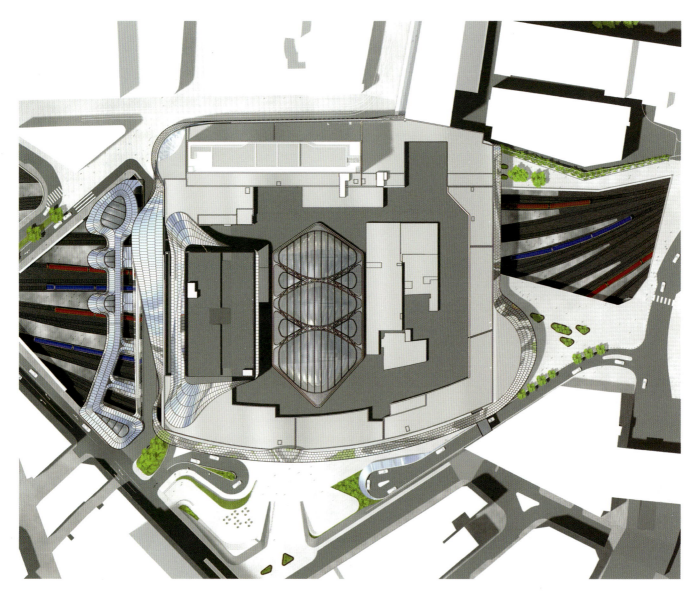

6.10 Roof plan, Birmingham New Street Station refurbishment by Foreign Office Architects. (Courtesy of FOA)

movement permeability and to help open up the economic potential of adjacent sites. Forecasts predict that the station regeneration will lead to 10,000 new jobs through the better utilisation of sites around the station.

The striking design is based upon expressing the fluidity of trains and the bifurcating pattern of rails. FOA have employed these geometries to inspire a design that conveys the historic character of the site as a transportation hub. The cladding in stainless steel reflects the external cityscape and sky with its warped surface, producing a distinctive identity for the building from many angles. One aim of the architectural language employed is to represent Birmingham's role as a modern industrial centre and to allude to the city's heritage of metal and jewellery crafts.

Beijing South Station

Designed by Terry Farrell and Partners, the new interchange serving southern Beijing is designed today to cater for 80 million passengers a year and, in 2030, for an anticipated 105 million. China's Ministry of Railways has described the project as an exemplar design that will be used to measure other planned transport interchanges. Key features of the project are the generation of both an architectural plan and an urban one by the same architectural practice, thereby securing integration between the interests of both; the creation of vertical movement zones to complement the horizontal ones; and the decision to locate the new interchange half a kilometre away from Beijing's historic main railway station (Wong, 2008: 108–10).

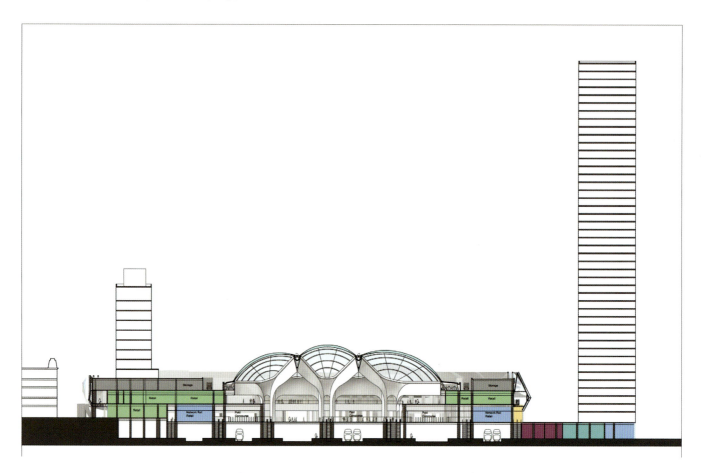

6.11 Section, Birmingham New Street Station by Foreign Office
Architects. (Courtesy of FOA)

6.12 Birmingham New Street Station, south west image. Design
by Foreign Office Architects. (Courtesy of FOA)

6.13 Interior of Beijing South Station by The Farrell Partnership. (Photo: courtesy of Zhou Ruogu Architectural Photography)

6.14 Interior of Beijing South Station by The Farrell Partnership. Notice the height necessary for natural ventilation and the incorporation of daylight shafts parallel to the tracks. (Photo: courtesy of Zhou Ruogu Architectural Photography)

As such, the project looks forward to an age of high-speed rail connection (rather than air) to the major cities of Shanghai and Tianjin, as well as the use of rail projects to revitalise older urban areas in Chinese cities.

Opened in 2008, the project was a key feature of investment for the Olympic Games. Like Shanghai South Station, the design is based upon a series of circles and crescents that allow the massive interior to be lit from above. Each crescent is tilted upwards towards the centre, thereby letting light and natural ventilation impact upon the volumes below. To aid orientation and to link two adjacent parks that act as gathering points for the interchange as well as gateways to the city, there is a central concourse nearly 200 metres wide and 350 long, which is lit from above by a 30,000 square-metre skylight. The scale is Piranese-like and supports

perception as much as it clarifies major and minor concourse routes.

The interchange puts pedestrian needs first, with taxis, buses and cars circulating around the perimeter. Passengers pass into the main concourse where tickets and travel information are obtained and then proceed downwards to the platform level. All trains, whether high speed, intercity or local, are placed at the same level, resulting in a huge expanse of track and

6.15 Beijing South Station by The Farrell Partnership. Notice the subtle references to traditional Chinese architecture in the profiling of the roof. (Photo: courtesy of Fu Xing)

platforms. To provide orientation and further support the environmental strategy, a secondary line of roof lights runs immediately above the line of the platforms.

Of particular interest is the way arriving and departing passengers use different levels in the manner of an airport terminal. This helps avoid movement conflicts between the planned 30,800 passengers an hour and helps enhance security. Vertical movement is via banks of lifts and escalators that help define the limits of the central concourse. Secondary vertical movement zones are located beneath the skylights of each crescent. Hence, the uplifting edges of the crescents establish the areas for passenger exchange between levels and also the zones for emergency escape in the event of fire or terrorist attack.

Another innovation in this important project is the use of extensive areas of photovoltaic panels in the roof. The 15,000 metres of solar panel provide electricity for the interchange and export surplus to the grid (Wong, 2008: 108–10). As China is now the world's leading manufacturer of PV panels, the station uses its expansive roof as a test bed for the new technology.

Shanghai South Station

The large Shanghai South Station provides 60,000 square metres of enclosed transport facilities beneath a circular, stepped roof. Designed by the French transport architects

AREP in association with East China Architectural Design and Research Institute, with engineers MaP3, the station provides an interchange for those people connecting between long-distance, regional and local rail services, as well as bus, taxi and underground metro systems. If mass transit facilities are one of the defining characteristics of the contemporary city, then this massive and dramatic interchange in the Xuhui district of Shanghai is one of the emblems of supermodernity.

The interchange is elevated about 40 metres above the ground, thereby providing an airy enclosure beneath the circular glazed roof. This is nearly 300 metres in diameter and consists of three layers of glass and polycarbonate with extensive aluminium sunshading. The filtered light bathes the concourses beneath in soft luminance reminiscent of the traditional shaded Chinese garden. The roof, shaped like a giant bicycle wheel, rises to an opening in the centre, which provides some of the cooling and ventilation for the interchange. It is supported by a perimeter ring of 18 huge steel columns fabricated by a local shipyard. Besides the concourses that take passengers to the web of facilities, there is a central waiting room with capacity to seat 10,000 people. Here, passengers can watch the movement of trains on the decks beneath. Like an airport terminal, passengers arrive at low level and depart at high level, boarding the trains via platforms at mid level. As such, Shanghai South Station provides a useful model of how to achieve high organisational standards with much visibility of facilities, while also providing good environmental conditions for passengers.

6.16 Section, Shanghai South Station, as designed by AREP. (Courtesy of AREP)

6.17 Interior view, Shanghai South Station, as designed by AREP. (Courtesy of AREP/T. Chapius)

Around the edge of the circular space the usual array of retail and leisure services helps mark the routes out of the interchange into the two major squares (north and south) that join the station to external commercial areas.

The interchange has an impressive scale and high level of ambition. Catering for over one million passenger movements per day, it is probably the largest single rail-based interchange constructed to date (other larger examples, for instance in Tokyo, are amalgams of earlier infrastructure and station provision). Shanghai South Station seeks to transform relatively inefficient urban infrastructure into a more sustainable model. It does this by connecting in an attractive fashion various modes of travel under a single roof, while also utilising the external area for pedestrian, cycle and bus interchange. The roof itself is a model of environmental and structural engineering that should be better known. Finally, it signals through the elegant, shallow, cone-shaped glass roof the importance of public transportation in a city previously committed to car ownership and struggling as a consequence to deal with heavily polluted urban air. In this sense, Shanghai South Station is a symbol of the shift in emphasis in China from private to public transportation as the country grapples with the demands of sustainable development.

Parramatta Transport Interchange, Sydney

This interchange is built around Parramatta Station, the oldest surviving railway station in Australia and hence listed as a national monument. The railway line was constructed in 1855 to link Parramatta with downtown Sydney, about ten miles away. Today it operates as a major suburban interchange with intercity and local train services and new bus provision. The plan by Hassell Architects is straightforward: a large wing has been added to one side of the existing station, which now acts as the interchange entrance, the buses have a dedicated area along the front of the new wing, and an underground passage links the whole to the nearby shopping streets.

Although the plan for the interchange sounds straightforward, the non-orthogonal arrangement of existing streets and a change in level across the site added complexity to the arrangement. This led to more steps and ramps than ideal, a

situation made more difficult by the raised pavements required for disabled access to buses (Thalis, 2007). As a result, the front of the station lacks the smooth transition from city streets to interchange concourse found elsewhere. However, the irregular plan geometry allows for taxis and cars to access the back of the interchange alongside widened pavements at this point, giving more modal choice to travellers. Since the railway tracks and platforms are at high level, much of the street traffic on foot and by wheel is taken through tunnels and along the sides of embankments. This has the effect of reducing the severance common to urban areas crossed by railway tracks, and here the tunnels and surrounding walls are enlivened by art works that extend for over 100 metres. With skilful use of lighting, the art works add richness to what otherwise would have been a blank wall.

The interchange is a tall, highly glazed, rectangular box, which contains the gantry for the overhead lines, protects the curving platforms, stairs and escalators, and extends to reach the edges of the pavements on either side. Considerable attention is paid to environmental features: the canopy has two major roof cut-outs to allow for ventilation, the perimeter walls are glazed from floor to ceiling to maximise daylight in the interchange, there is natural cross-ventilation and, finally, there are external fixed louvres that reduce glare inside and shade the façades from direct sunlight. The contrast between the transparent new interchange and the solid classical station of 1855 says a great deal about current attention to the comfort and amenity of passengers rather than to image, which was the concern of the original railway company (ibid.).

Linkage, both physically and commercially, is an important consideration in the design of interchanges. Here, escalators and ramps take people down to an enlarged shopping concourse

6.18 Roof view, Southern Cross Station, Melbourne, designed by Grimshaw. (Photo: courtesy of Shannon McGrath)

6.19 Interior of platform area, Southern Cross Station, Melbourne. (Photo: courtesy of John Gollings)

beneath the station at its centre and to a smaller concourse at the east end. The concourses connect with adjoining streets via underpasses and malls and compensate for the restricted street-level access, which stems from the decision to locate the bus interchange immediately alongside the station. Since pavements are kept dry by the overhanging canopy of the main concourse, the steps, ramps and escalators do not pose such a hazard had they been exposed to Sydney's rains.

Southern Cross Station, Melbourne

Southern Cross Station in Melbourne, designed by Grimshaw in collaboration with local practice Jackson Architecture, is a major bridge between Melbourne's regenerated docklands and the centre of commercial life in the city (Raisbeck, 2007). The building occupies a large city block, with the trains coming in at low level and passengers and buses at street level. The 16 tracks serving 8 main platforms are slightly angled to the street grid, allowing the bus interchange to be tucked into the acute angle of Spencer Street, which forms the main perimeter street to the station. Since the former railway tracks and docklands formed a rather neglected industrial quarter of Melbourne, one major design task was to reclaim the area through massive infrastructure investment.

The new station fills the city block with concourses, booking halls and platforms. A high-level bridge takes pedestrians across the tracks to a new shopping area at the north-east of the station. There are no cross-routes through the station beyond that of the two main perimeter malls at the Collins and Spencer Street façades. As a result, the station has a traditional arrangement of railway-focused activities, which look out on to the city through lofty, glazed façades.

As with Grimshaw's transport architecture elsewhere, the roof is a major element both in the experience of internal volumes and in landmarking the interchange within the urban field. As Sir Nicholas Grimshaw notes, 'getting daylight into

transport buildings is important for orientation but direction of structure matters – it is important not to go against the grain of movement' (Grimshaw, 2009).

The station roof at Melbourne's Southern Cross Station mirrors three of Grimshaw's concerns: first, that the roof is attractive when looking down on it from surrounding high buildings; second, that it is uncluttered and appears to float above the platforms; and, third, that it signals the major functional zones and circulation routes (ibid.). These principles are necessarily adjusted to site contexts – aesthetic, spatial and climatic. However, they provide a clue to the attention paid to the roof in the architect's transport buildings, from Waterloo to the Fulton Street hub in New York and here in Melbourne.

The configuration of the direction of the railway tracks, the shape of the city block and the external links to buses means that the roof at Melbourne converges from four main undulations to three. The roof follows the line of platforms rather than the geometry of perimeter roads, thereby signalling the importance of the train over the car. However, it expands at the Spencer Street side to accommodate the bus connection to the station. These clever design moves give this complex transport interchange its legibility, especially as the roof has a line of ETFE pillows that follow the platforms below.

The roof is rather more an undulating canopy of 'dune-like moguls', which reflect the deformations made by the wind and internal air movements (Raisbeck, 2007). As such, the roof has a living quality that alludes to Grimshaw's particular interest in ecological architecture. Daylight and filtered sunlight enters the building via the roof, where the wavy spine of ETFE pillows sits above Y-shaped columns that, in turn, march down the centre of the platforms. The undulations and the perimeter glazing give the station roof a floating impression, and also provide enough volume beneath to accommodate the various sub-structures without the loss of legibility.

Two main concourse spaces link the station to the city. The first joins the important Collins and Spencer Street corner to the station booking hall, ticket office and waiting lounges. The second links together the bus stop areas to the station areas with a long concourse parallel to Spencer Street. This wedge-shaped space has its own ticket area and a number of double-height retail spaces. Here, the undulating roof rises higher than elsewhere to signal the location of the commercial zone.

Bijlmer Station, Amsterdam

Bijlmer Station is a large multi-track interchange in the suburbs of Amsterdam, serving the 50,000 capacity stadium known as ArenaA (the home of Ajax football club) and the wider community. It consists of an elevated interchange between train, bus and metro services, which is stitched into a 70 metre-wide pedestrian boulevard that leads in one direction to the stadium and in the other to commercial and housing areas. Designed by Grimshaw in collaboration with the Dutch practice Arcadis Architects, the project is marked by an ambition to celebrate public transport as the primary gateway to sporting arenas.

The facilities are large and ambitious. The interchange consists of a ticketing and retail space of 6,500 square metres, with platforms 340 metres long for high-speed trains and 150 metres long for suburban rail. The cost, including external

landscape, was 130 million euros, making Bijlmer one of the most important rail interchanges built to date.

The geometry of the interchange is based upon right-angled connections with the station at high level and bus, metro and pedestrian facilities at ground level. In order to enhance the scale and drama of the interchange, and to provide generously proportioned spaces, the train platforms were raised 2 metres above the original proposals in the masterplan for the area. The benefits are twofold: first, the area beneath the station becomes more welcoming, adding to the sense of safety and security; second, by elevating the station, it serves better as a landmark in an area of the city where visual cues are important and where visiting football supporters need help with orientation.

The interchange is designed to handle 60,000 passengers a day and many more on match days. Hence, the station spaces are generously proportioned and link well into the pedestrian boulevard designed by Architekten Cie. The physical

6.20 Bijlmer Station, Amsterdam, designed by Grimshaw in collaboration with Arcadis Architects. (Photo: courtesy of Ger Van der Vlugt)

layout expresses in a direct way the fusion of interests and financial inputs from the two main participants in the project – the Dutch government organisation in charge of rail infrastructure, known as ProRail, and the City of Amsterdam planning department, representing community interests (Kolb, 2008). As with many transport projects, the practice of Grimshaw had the task of liaising with many other design and engineering consultants in bringing this project, shortlisted for the Stirling Prize in 2008, to fruition.

The circulation of buses and metro on the ground floor helped determine the spacing and size of the columns that support the eight-track station above. The concrete columns are massive and help direct pedestrian movement to the station entrance, which is marked by a glazed screen. The columns are rounded and spread above head level to take the wide loads of the railway tracks above. In order to avoid graffiti, they are covered in easily cleaned mosaic tiles which, along with the finely engineered detailing elsewhere, add quality to the undercroft spaces. As Sir Nicholas Grimshaw notes, 'people expect higher finishes outside the UK' and, although structural design matters, 'good signage, good lighting, good finishes and good loos (toilets) are what often matter most to passengers' (Grimshaw, 2009).

Within the cityscape the Bijlmer Station Interchange is marked by the folded patterns of the glass and steel roof of the platforms and the perimeter skirt of louvred solar screens. The effect is to produce a distinctive play of sunlight, shadows, reflected light, diagonal views and architectonic effects. Light and associated shadow patterns animate the platform areas and the bus circulation space below. In typical Grimshaw fashion, the primary routes through the interchange are marked by elegant steel and concrete structures. These, in turn, lead to dramatic circulation volumes that are more spacious than is usual and recall the sense of arrival at Waterloo International. To soften the aesthetic language and to help with acoustics, the underside of the platform canopies are lined in pine – the wood giving an unexpected air of quality to waiting spaces.

The roof is a pattern of angled glass, stainless steel and aluminium. Each pair of platforms is divided by an A-shaped, double-height column, which is glazed on both sides immediately above the platform. The five lines of columns provide the primary spatial order for the station and, in turn,

define the vertical circulation routes and the geometry of the platforms and, as a result, give the main means of orientation for the passengers. Though the interchange is complex, the section and plan have a logic that runs through from detail to the macro-planning of the area. As William Curtis notes, out of the urban chaos of suburban Amsterdam comes a 'hybrid building with spaces of real quality bathed in daylight coming through the crowning element, the oversailing roof' (Curtis, 2008).

Three German examples

Germany has invested heavily in high-speed rail transport (notably ICE) and, as a consequence, is restructuring many of its major railway stations. These have been transformed from mainly singular provision into multi-modal transport centres. Two characteristics of rail investment in Germany stand out. The first concerns the use of high-speed trains to help unify Germany after the fall of the Berlin Wall in 1989. Stations such as the new Berlin Central (which caters for 350,000 passengers per day) almost straddle the divided city and other station remodelling, such as at Leipzig in the former East Germany, help to bring the nation's people together again both physically and spiritually.

ICE trains are the German equivalent of the French TGV system and are capable of travelling at nearly 200 miles per hour. Just as in France, ICE has led to a station renaissance and, in particular, to the growth of inter-modality. Railway stations are seen as points on networks where transfer to bus, tram, plane, metro, taxi, bike and foot occur (Niedenthal, 2008). Many of the new ICE stations are drastic remodellings of older facilities, but some are completely new, such as Berlin Central, while others are still planned, such as Stuttgart Hauptbahnhof. As in Britain, a number of fine stations were lost in the war or demolished in the 1960s, but many remain and these have been the centre for innovative expansion and sensitive restoration. Three of these railway stations, which involve conversion to interchanges, will be discussed briefly.

Dresden Central Station is one of the largest stations in Europe and has been extended and remodelled to designs by Foster + Partners. Like all high-speed train stations, most

6.21 Dresden Station, as redesigned by Foster + Partners.
(Courtesy of Foster + Partners)

passengers have already travelled by some other means by the time they arrive at Dresden Central. Hence, the emphasis is on transfer on foot and by wheel, and on ensuring that orientation is not compromised by excessive infilling of space. Foster's design retains and restores to its former glazed elegance the triple-vaulted roof, which provides cognitive legibility at the heart of the station. The aim was to strip away additions and alterations in order to restore the integrity of the original design and rationalise circulation (Foster, 2008). The team was careful to ensure that the listed elements read as historic, while the modern interventions were perceived as contemporary.

The original station is framed by two high-level wings of through trains and metro services, with a central area of terminating lines on the central axis at lower level. The split in section allows passengers to walk beneath the elevated tracks, where access at ground level is provided to connecting bus services. A new square created at the front of the station provides the major foot, cycle and taxi connections.

To accommodate the new ICE trains, the platforms were expanded and extended. Rather than repeat the structural arrangement of the original station roof constructed in 1898, Foster + Partners devised a translucent silver-white Teflon-coated membrane, which spans between 67 rigid, glazed ribs. The roof consists of three ribbed arches – a large one in the centre and smaller ones on either side – which cover the platforms and circulation areas and connect to the original concourses.

Leipzig Central Station is another expansion to accommodate ICE trains. Again, there are listed elements and extensive new facilities to cater for commuter and transfer passengers. The station is large and highly glazed, and covers 84,000 square metres (Niedenthal, 2008: 34). The old station concourse has been converted to a three-storey shopping mall, with passengers passing through at low level in the manner of Atocha Station in Madrid. The new extensions to the station consolidate services that previously ran behind the shopping mall, which now extends the full width of its 26 platforms. A

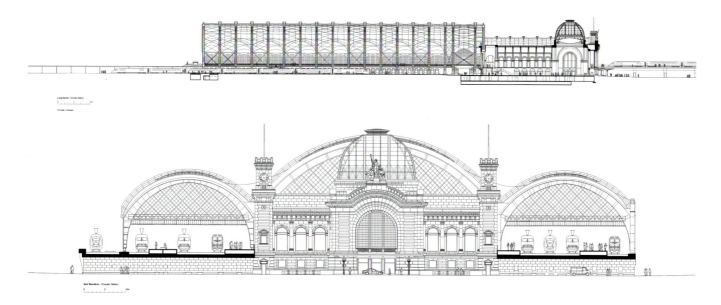

low-level transport tunnel planned in the future for metro services will take underground movements at right angles to the tracks and hence to the city beyond. The main lesson from Leipzig is the hybridisation of transport and retail functions within an extensive existing building claimed to be the largest railway station in Europe. Culturally, the handsome old station has been retained and its dimensions extended to serve transport needs at one end and commercial ones at the other. The new station complements the old in scale and technological ambition. Old and new come together to form an interchange, while also providing a new retail hub for the city and a magnet for wider urban regeneration.

Berlin Central Station, which opened in 2006, is located where once the Berlin Wall stood. It takes advantage of land set aside for security and political demarcation. Today it marks the spirit of reunification and is the centre of an ambitious wider programme of urban regeneration. As such, the station, designed by the Hamburg practice of von Gerkan, Marg and Partner, is planned to act as an urban dynamo within one of Berlin's few underdeveloped central areas. However, like all land around major transport infrastructure, there are many barriers to movement, including a busy highway and not far away the curving river Spree. One of the objectives was to create an interchange with access points in all directions. Another was to entice car drivers to use the new station as a park and ride facility, employing an existing car park alongside the station. The design of the new station consists mainly of steel and glass platform roofs that span both platform and concourse areas, inspired perhaps by Grimshaw's example at Waterloo International. The arrangement allows the station to be one open hall with cross axes in the manner of market or exhibition buildings. It is perhaps the nearest any modern station has come to interpreting the Crystal Palace for the railway age.

Interchange occurs at various levels via connecting lifts and escalators. Since the station is highly transparent and open in nature, views across the massive internal volumes allow

6.22 Long section, Dresden Station. (Courtesy of Foster + Partners)

6.23 Cross-section, Dresden Station. (Courtesy of Foster + Partners)

6.24 Concourse, Berlin Central Station. (Photo: courtesy of Odilo Schoch)

6.25 Curving structure of Berlin Central Station. (Photo: courtesy of Odilo Schoch)

passengers to navigate between high-speed train, local train and metro services. The high-speed rail tracks enter at high level, with local rail and metro service beneath. Connections to bus, taxi and car occur around the periphery. An arched canopy signals across the river to the government offices opposite and provides a point of entry to celebrate one of Germany's busiest railway stations.

Four French examples

Just as ICE has led to the conversion of some notable existing railway stations in Germany into interchanges, the same is true in France under the impact of TGV rail investment. However, although the pattern in France (as at Lille Europe Station) is rather more the creation of new interchanges than

the remodelling of existing facilities into multi-modal transport hubs, there are still important lessons to be gained from French stations. Several examples are worthy of note, including the new interchanges at Valence TGV Station and Avignon TGV Station, and the restructuring of Gare du Nord in Paris and Strasbourg Station in eastern France. All four were undertaken by AREP, the large Paris-based architectural and engineering practice that has done much to revitalise transport architecture since its inception in 1997. AREP is a subsidiary of the French national railway company SNCF, but now provides designs around the world (it was, for instance, involved in the Beijing and Shanghai South transport interchanges).

The Valence TGV Interchange connects the high-speed rail network with regional trains and buses in the Rhône Valley in central southern France. The new station is where the rail lines cross and, hence, infrastructure considerations are the main

6.26 Lille Europe Station. (Photo: Brian Edwards)

determinant of location. The site is semi-rural and this has encouraged the design to adopt extensive park and ride facilities as well as bus connections. The TGV line runs at an angle in a cutting 8 metres beneath the surface-located regional train network. Passengers move between the two in an elevated glazed concourse, with stairs, lifts and escalators dropping down to the two train levels. Bridges and decks extend outwards to and from connections with car and bus services, and open up extensive views over the surrounding landscape. They also provide views down on to the two platform decks, thereby giving users a good grasp of the interchange layout.

The design is conceived as a glazed nave, which floats above the landscape, providing the appearance, according to the architects, of a building levitating above the tracks (Browne, 2008: 107). The interchange is long at 320 metres (roughly the length of the TGV trains) and, being steel framed and clad in glass, has an inviting transparent quality. Floors are decked in timber and the roof is lined in plywood, giving the building a natural character in keeping with the Rhône Valley. The cross-section is arguably more interesting than the plan, with its over-

sailing roofs (to limit solar heat gains), battered retaining walls (to deflect noise) and split Y-shaped steel frames (to animate the movement rhythms for passengers passing in trains). As a result, the building functions well and sits elegantly within a landscape noted for its vineyards and tourism.

Avignon TGV Station lies just outside the ancient city of Avignon in Provence. Designed by AREP, the station follows the curve of the high-speed railway tracks, themselves dictated by the geometry of the River Durance. It is an interchange by virtue of the connecting bus and taxi services that run on the north side of the station. Like Valence TGV Station, there are large areas of park and ride facilities, which are an inevitable result of the semi-rural location of Avignon TGV Station. Two points are of particular note. First, the station responds in a dynamic way to the local climate – shading itself from the heat of the sun and protecting itself from the mistral winds while also exploiting breezes for cross-ventilation. The dynamics between sun and wind lead directly to the cross-sectional arrangement with its curves and overhangs, and areas of glazed and solid wall (ibid.).

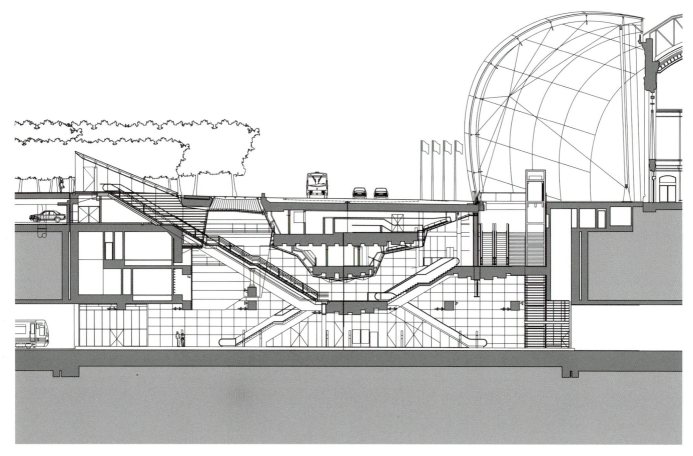

6.27 Cross-section of public concourse at Strasbourg TGV Station, designed by AREP. (Courtesy of AREP)

The second innovation concerns the use of airport termin-ology and planning. There is an arrival hall and corresponding departure hall with the railway tracks between. Passengers move between the two in a tunnel under the tracks. The arrival hall, where tickets, train and bus information is obtained, is a relatively simple structure compared to the departure hall. This is curved in plan and section and shaped to exclude the sun while opening up views to the north. The curved plan, figured like a segment of an orange, allows the departure hall to offer little resistance to the mistral winds. Passengers wait in this airy well-lit space for the arrival of the train rather than stand outside on the platform, which is the normal practice. In this sense, the arrangement follows airport practice, as do the limitations imposed by the choice of location.

Gare du Nord in Paris has been remodelled to accommodate both TGV and Eurostar services, again to designs by AREP. Like their London equivalents, the major Paris stations were already complex interchanges before the advent of high-speed trains. The project involved joining together two existing stations (the historic Gare du Nord, designed by Jean Hittorf in 1861, and the more modern Magenta Station carrying regional trains)

by large glazed concourses that connect in turn to metro lines. Like London's St Pancras, the complexity of interchange is enormous, particularly since many pre-existing infrastructure arrangements had to be respected.

The three main transport modes of international high-speed train, regional train and metro services are connected via a combination of glazed and underground walkways, and large cavernous volumes where escalators and lifts are located. Above ground, the new architecture follows the nineteenth-century tradition of skeletal steel frames and glass while, below ground and in the passageways, AREP have carved out dramatic volumes open to the sky above. These allow views upwards to the Parisian street scene, thereby helping to provide orientation. One feature of the design is the way the footprint and proportions of the historic buildings have been retained and reinterpreted to ensure a degree of cultural connection between new and old (ibid.: 99).

Strasbourg Interchange serves the eastern arm of the TGV network and operates as a gateway to German high-speed rail infrastructure. The original Strasbourg Station has been doubled in size to allow for the convergence of TGV and regional rail

6.28 Forecourt as gathering space and public park at Strasbourg TGV Station, designed by AREP. (Courtesy of AREP)

services, with buses and taxis using the adjoining station square and new metro threaded underneath. A key element of the new arrangement is the creation of a large curved, glazed concourse which, like the skeleton of a whale, now fronts the old station. TGV has radically altered the functional and aesthetic landscape of this historic station.

The new concourse, which fronts the old façade, acts as a gathering point for all passengers. From here you can progress to the various train services beyond, pass through to the bus stands in the square or travel down to the metro services underneath. The square itself is a major urban space for a variety of civic as well as transport-related functions. It has been dramatically remodelled from its nineteenth-century origins to form a gateway space to modern Strasbourg. In fact, the relationship between external space and internal concourses is particularly well orchestrated in terms of three-dimensional passenger movement.

Strasbourg works well as an interchange and demonstrates many of the principles outlined by Fiona Scott in *InterchangeABLE* (see pages 55–6). Attention has been paid to ribbons of connection between the city and the transport infrastructure, movement is barrier free, the information points explain the linking services and are cross-timetabled, and landscape, urban and building design are seamlessly integrated. Also, rather than obscure the historic station façade, the new glazed concourse opens it up to close examination via a climate-protected walkway.

Four interchanges in the USA

Without the high-speed rail infrastructure investment that characterises transport interchanges in Europe, inter-modal stations in the USA are generally more modest in scale. These consist generally of two types: first, upgrading of existing facilities to provide new or improved interconnection and, second, extension to existing infrastructure. Since transport architecture takes a long time to be realised, the political ambition of the Obama administration to revive railways and invest in high-speed rail has yet to lead to any new interchanges. However, much has been achieved at a more local state and city level, often motivated by concerns over urban air quality, social inclusion and the needs of the elderly.

A typical example is Roosevelt Avenue Inter-modal Station in Queens, New York, designed by architects Fox and Fowle. It consists of a transfer station between elevated and underground metro services and buses to LaGuardia Airport. The interchange is triangular in plan, with retail buildings at the apex and a lofty, glazed concourse arching across the centre between buses on one side and rail services on the other. Steel framed and clad in blue-green terracotta panels, and extensively glazed to the north, 'the building acts as a modern beacon for the diverse neighbourhood' (Wortman, 2005: 67). Of particular interest is the way environmental factors have shaped the design, giving the interchange strong civic presence. For example, the tall, glazed concourse acts as a natural ventilating chimney for the subway – the shape is designed to draw the warmed air out of the low-level platforms using the stack effect. Louvred windows and roof vents at high level in the concourses exhaust the air, which is replaced by fresh air taken in at low level through filters. The curved profile of the roof, needed to encourage thermal air movement within the concourse, supports photovoltaic panels on its south side. These provide much of the electrical energy to run the lighting below ground and also fans needed to circulate the air on still, sunless days. A combination of maximised daylight, natural ventilation and new energy technologies has resulted in a distinctive building that serves as a new community focus in Queens. In many ways, the quality of the design grows directly out of the New York City Transit Authority's goals of environmental responsibility, high-performance design, neighbourhood integration and ease of maintenance (ibid.: 69). This highlights the importance of an enlightened, broadly based transport brief in producing good civic architecture.

Another project by the same architects is Hoboken Interchange in New Jersey, which combines light rail, heavy rail and ferry services into a loosely composed grouping. Fox and Fowle were mainly responsible for the new light rail station on the Hudson-Bergen Rail Transit System. The site is on the historic waterfront of Hoboken and provides smooth transfer between ferry, rail and light rail services by opening up a former private area of harbourside for public use. Both rail and ferries terminate at the site, with the new light rail station providing 'dynamism and balance' to the pre-existing arrangements (ibid.: 70). The structure of the light rail station is light and airy, using the language of traditional station architecture to forge harmony between new and old. Because of the nature of the site, curved glass windscreens were provided to shelter passengers waiting at the elevated platforms. As with Fox and Fowle's other projects, attention to sustainable design practices and to incorporating the work of local artists sets the Hoboken Inter-modal Station apart from many others.

Fulton Street Transit Center, New York

The vast underground Fulton Street Transit Center, masterplanned and engineered by Arup with the above-ground station elements designed by Grimshaw, is planned to handle nearly 300,000 passengers a day. It is part of the huge exercise in urban reconstruction and transport improvements planned to be undertaken by the New York Metropolitan Transit Authority. The Arup–Grimshaw Fulton Street hub is planned to link in with an above-ground interchange three blocks away, designed by Santiago Calatrava as part of the rebuilding of the World Trade Center site. The project as a whole is a valuable lesson in accommodating both the scale and complexity of new transport buildings within mature cities and in turning transit infrastructure into architectural landmarks. The Fulton Street Transit Center, in particular, is a lesson in how to modify existing transport provision to serve the needs of the twenty-first century, putting in place those things such as environmental quality, light, space and legibility that were lacking in the original arrangement. One characteristic of American public transport, particularly in the big cities, is the lack of connection between the parts, which

6.29 Section, Fulton Street Transit Center, New York, designed by Grimshaw. (Courtesy of Grimshaw)

6.30 Interior image, Fulton Street Transit Center, New York, designed by Grimshaw. (Courtesy of Grimshaw)

is a legacy of lines built by competing companies. Today, designers and engineers have the task of better joining up the fragmented and often neglected pieces of the public transport jigsaw. President Obama's Stimulus Funding package of 2010 is partly aimed at addressing this problem.

The design by Grimshaw for the head house element (the main above-ground interchange space) of the Fulton Street Transit Center is within a development framework established by Arup. Grimshaw's team, working in collaboration with artist James Carpenter (using funding under the 'Arts for Transit' scheme), has designed a 35 metre-high truncated, stainless steel cone with a glass oculus in the centre. This sits above a generously proportioned booking hall that leads passengers to the subterranean railway lines. The role of the glazed oculus, which is 55 feet in diameter and incorporates prismatic elements, is to channel and reflect light into the existing network of underground railway and metro systems. To amplify light levels, a net of metal panels of varying specularity (or reflection) is employed. From the outside the effect will be to form a landmark and point of orientation in an area of skyscrapers. At night, it will shine like a beacon in the surrounding streets of nearby Broadway in central New York.

The whole station, with its transparency and sense of theatre, helps signal rebirth and counterbalances the skeletal structure of Santiago Calatrava's larger transport interchange planned for nearby. It also makes a contrast to the more solid architecture of the vicinity, such as the landmark terracotta Corbin Building. The ticket office and main concourse will contain cafés and shops arranged mainly as a gallery at high level. Hence, the new Fulton Street hub will be a place to visit, to meet friends and to shop, while also catering for the transport needs of urban New Yorkers.

The Fulton Street Transit Center is a stylish and sophisticated landmark that uses high-quality materials and generous circulation spaces to signal the importance of low-carbon travel. The design concept is simple – a semi-transparent cone set within a cube of glass recalling, perhaps, the play of sculptural forms in the work of Le Corbusier. All is visible, including the routes down to the six underground railway stations and, in the opposite direction, the streets outside. The involvement of Carpenter gives the project local artistic interest,

6.31 Congestion is often severe when historic stations are subject to tourist pressure. Here, tour buses prevent Toledo Station in Spain operating as an effective interchange for local people. (Photo: Brian Edwards)

thereby helping to give the design a cultural as well as architectural and engineering dimension.

The Fulton Street site is not far from the former twin towers of the World Trade Center and, hence, has sensitivity from many points of view. Here, the aim is to help regenerate the area, to connect the businesses of Lower Manhattan to improved public transport, and to provide a dignified route to the proposed World Trade Center Memorial. As such, the cone of light is both a signifier and a beacon of hope to many Americans.

Table 6.1 Train interchanges

Name	Type	Architect	Characteristics
Kashiwanoha Campus Station, Japan	Train/tram	Makoto Sei Watanabe	• Sleek form • Solar ventilation
Kashiwa-Tanaka Station, Japan	Train/bus	Makoto Sei Watanabe	• Pod-like shape • Bridge station
Moscow Central Station	Train/metro/bus	Behnisch	• Air rights exploited • Many green design features
St Pancras Station, London	Eurostar/intercity/metro/bus	Foster + Partners; Allies and Morrison; John McAslan + Partners	• Utilises existing historic structures • Urban space reconstruction
Leuven Station, Switzerland	Train/bus/cycle	Samyn and Partners	• Incorporates cycle bridge • Existing station retained
Shanghai South Station	Train/metro/bus	AREP with East China Architectural Design and Research Institute	• Multi-level interchange • Solar protection • Highly glazed • Circular form
Beijing South Station	Train/bus/tram	Farrell Partnership	• Circular form • Alludes to traditional Chinese architecture • Many green features
Arnhem Station, Netherlands	Train/bus/trolley bus	van Berkel with Ove Arup	• Dynamic form based on fluid flows • Multi-level
Melbourne South Station	High-speed train/local train/bus	Grimshaw	• Wavy roof • Highly glazed • Economic link
Florence Station (project)	High-speed train/local train/bus	Foster + Partners	• Highly glazed • New parallel platforms • New urban space
Parramatta Interchange, Sydney	High-speed train/local train/bus	Hassell	• Listed station extended • High environmental standards
Gare du Nord, Paris	High-speed train/regional train/metro/bus	AREP	• Listed station extended • Links together two stations • Large glazed concourses
Strasbourg Interchange	High-speed train/regional train/metro/bus	AREP	• Listed station refurbished • New curved concourse at front • Square for metro and bus connections
Bijlmer Station, Amsterdam	Train/bus/cycle	Grimshaw	• Access station for Ajax Stadium • Dynamic forms • New urban axis created
Dresden Central Station	Train/bus/metro	Foster + Partners	• New civic space created • Extension to existing platforms • Restored glazed areas
Berlin Central Station	ICE train/regional train/bus/metro	von Gerkan, Marg and Partner	• Built on Berlin Wall • Large concourse retail areas • Bridges the divided city

Ferry interchanges

The ferry interchange differs from other types in four important ways. First, normally the flows of people are greater in smaller time slots, thereby putting facilities under stress for relatively short periods of time. Between the arrival and departure of ferries, the interchanges can be quiet and leisurely places to watch ships passing or have a coffee. Second, the highly visible movement of ferry boats gives ferry terminals an air of expectancy and openness to both the water and to the means of transport. Third, the linking network of bus and train services is always to the landward side, where roads and railway tracks invariably have to be crossed. This results in many ferry interchanges having high-level foot bridges and complex movement patterns. Finally, the ferry interchange (like the airport) is frequently located at a national border and, hence, customs and immigration controls may have to be conducted. So, although within the taxonomy of types the ferry interchange shares features with bus and train interchanges, there are distinct differences that designers should be aware of and could possibly exploit.

Many major cities of the world, such as New York, Sydney and Hong Kong, depend heavily on boat-based urban transportation. Thus, ferry terminals are part of the fabric of these cities and contribute greatly to their character and business efficiency. The ferries transport millions of blue- and white-collar workers from their homes in the suburbs, often by a web of linking bus and rail services. Here, there is not usually one terminal but a collection of terminals serving different destinations and often run by different companies, sometimes on a highly competitive basis. However, in New York's case, the city transportation authority runs a fleet of ferry boats operating out of five terminals, and a similar pattern is found in Vancouver and Seattle. Normally, in the west, ferry transport is a civic venture, while in eastern countries it remains largely a matter of private enterprise. The distinction affects the design ethos of ferry interchanges as well as the quality of the facilities that are available. As many ferry terminals were constructed in the 1930s and 1950s, there has been much recent reconstruction and design of new facilities. These have often been related to wider urban regeneration of harbour areas (e.g. Baltimore, Southampton and Rotterdam) and have been subject to conversion from singular terminals into transport interchanges. As such, the trend is towards better external linkage via pedestrianised routes and cycle ways, and the physical connection to local rail and bus services.

Ferry interchanges are part of the altered face of urban waterfronts. The phenomena of dockland regeneration and urban restructuring that characterised the latter decades of the twentieth century left their mark on ferry services. Boat-based transport emerged as a viable alternative to other forms of public transportation, especially where new economic or civic facilities were built near waterfronts. Older, established ferry services were augmented by new provision, often involving inter-modal connection. So the concept of ferry terminals being inter-modal centres became established and subtly altered the type. Added to this, it is now recognised that water-based transport emits less CO_2 than other public means of movement. The relative energy efficiency of transport by water is encouraging a reassessment of older ferry provision in the interests of sustainable development.

Key principles to follow in the design of ferry interchanges are:

• Maintain transparency and visibility of the water.
• Provide waiting areas overlooking the water.
• Provide adequate pedestrian space for movement around the terminal.
• Put people's needs before those of the connecting transport infrastructure (bus and rail services).
• Design the ferry interchange as a water gateway to the city.
• Link design to the waterfront context in terms of architectural language.

In architectural terms, many new ferry terminals have adopted the sleek lines of the cruise liner, others draw their inspiration from surrounding industrial structures, and others draw theirs from the functionalism and elegance of the modern ferry boats themselves. One great advantage of ferry interchanges is their location. Being by the waterside offers the kind of architectural potential denied of airports and bus stations. Water forms one of the great urban design edges in our modern congested cities, with the ferry terminal bridging the world of land and water. As such, it is a symbolic building that refers back to the tradition of water-based transport and forwards to the low-carbon age.

7.1 Typical harbourside with ferry terminals in Aarhus, Denmark. (Photo: Brian Edwards)

Since many mainline railway tracks follow the old docksides of industrial cities, ferry interchanges here often have to negotiate extensive barriers. A good example is Helsingor Interchange in Denmark, with its linking ferries to Helsingborg in Sweden and trains to Copenhagen. Ferry passengers have to cross the railway track and harbour road on a high-level glazed bridge, which provides at one end a link into the railway station and at the other an elevated boarding point on to waiting ferries. To make journeys easier there is good provision of lifts and escalators, and ticket facilities and timetabling information for the ferries and trains are on the bridge. Although the bridge is inconvenient, it offers good views of the ferries, trains and bus services while also providing a prospect across the harbour to the open water beyond. Since the ferries run frequently, their visibility allows passengers to time their journeys to the minute.

Cruise liner terminals are a distinct type. They require customs areas and connection to tourist coach services. Where they share facilities with commuter ferry services, as at Yokohama, they assume a scale and complexity similar to that of railway or bus interchanges. The presence of cruise ships adds prestige as well as scale, and this leads to raised architectural ambition. Tourists expect style as well as function, and space to experience new locations at the critical gateway between ship and land.

Yokohama Ferry Terminal

This is a large interchange for ferry and bus services and cruise liners in Yokohama, Japan, designed by Foreign Office Architects. What is most remarkable about the building is the way the roof acts as the medium of gathering and promenade, with the ticket offices and transport hall underneath. The roof is in fact an undulating deck in timber and grass, which provides the local community and tourists with a viewing point over the harbour area. Passengers board their ferries or liners at the perimeter of the large landscaped deck, often accompanied by musicians who use the timber slopes as an informal concert space. The complex, designed rather more as a landscaped peninsular than architecture, shows the power of

transport and water to entertain and provide a public spectacle out of passenger transfer.

Essentially, there are three levels: a lower service and parking floor, a central cavernous space for buying tickets and gaining information, and the landscaped decks above. Ferry boats use one side of the artificial peninsular, cruise liners the other, with local bus services reached at the landward end. Passengers are steered towards their boat or the city by a series of gently sloping floors, which also include customs points and waiting areas. Light enters via folds in the roof, cuts in the floor or directly from perimeter glazing. The effect is to break with traditional notions of terminal space, introducing the idea that fluidity and fragmentation add richness to transferring between transport modes. This is a ferry terminal where the internal spaces are carved out of the ground and the external ones rise and fall like sand dunes. The effect is one of blurring the distinction between interior and exterior spaces, international travellers and locals, public and private zones, and utility and pleasure. The timber decks act as a democratic space projecting into the sea, creating an active surface that challenges the symbolic gate-like quality of the building type (Jones, 2006: 140).

DFDS Ferry Terminal, Copenhagen

The long, sleek DFDS Ferry Terminal in Copenhagen, designed by 3XN architects, serves the ferry services to Oslo and to various destinations in the Baltic. Unlike those at many ferry terminals, the boats are large and carry both foot passengers and wheeled vehicles. This results in huge volumes of people passing through for short periods, and corresponding spacious waiting and customs areas. About 2,000 passengers use the facility daily.

The ferry terminal sits as a landmark in the harbour zone between land and sea. It is placed parallel to the water and sits elevated above the pier, where the building acts as a barrier between free and customs-controlled zones. Passengers rise at entry inside the building via stairs and escalators and are taken into a large linear ticketing hall with security and passport area beyond. The space is glazed in bands of transparent and

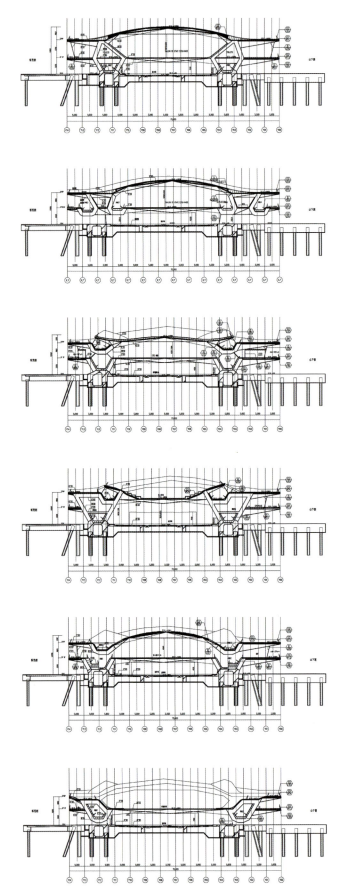

7.2 Sections, Yokohama Ferry Terminal, Japan, designed by Foreign Office Architects. (Courtesy of FOA)

7.3 View of Yokohama Ferry Terminal from across the harbour. (Photo: Brian Edwards)

7.4 Viewing and circulation deck, Yokohama Ferry Terminal. (Photo: Brian Edwards)

opaque glass, providing an attractive pattern of views out to waiting ferry boats. Since passenger and ticket facilities are at high level, the movement into the ferries is smooth and achieved with the minimum of inclines. Placing the public waiting areas at high level frees the ground floor to provide access for vehicle movements, as well as secure areas for the customs functions associated with ferries. Also, since the bus connection is at a right angle to the prow of the ferry terminal, the geometries of movement are logical and easily comprehended.

Being located in a port area between land and sea encouraged the development of a building that had a modern nautical character. The long, sleek lines and horizontal banding suggest a highly polished elevated container at the water's edge. Using different types of glass, the DFDS Terminal is designed to be a lighthouse by night, an important consideration for winter-night sailings. The façades are clear glazed at 40 per cent, this being the optimum for balance of solar gain, daylight penetration and cooling on the long south

7.5 DFDS Ferry Terminal, Copenhagen. (Photo: Brian Edwards)

7.6 Site plan, Ocean Terminal, Southampton, designed by The Manser Practice. (Courtesy of The Manser Practice)

7.7 View of the Ocean Terminal, Southampton, designed by The Manser Practice. (Photo: courtesy of Morley von Sternberg)

façade (Nielsen, 2009). The height and length of the terminal share dimensions roughly with those of the ships it serves and, being positioned between the two births, it provides a wind break for embarking passengers.

Ocean Terminal, Southampton

Designed for Associated British Ports by The Manser Practice, with detailed input from Stride Treglown, Ocean Terminal is a large interchange facility for cruise liners on the south coast of England. With an area of nearly 8,400 square metres, Ocean Terminal caters for the security, customs, health and transportation transfer needs of cruise ships with 5,000 passengers and 2,000 crew (The Manser Practice, 2007). As such, the terminal approaches in scale a medium-sized railway station. Spaces are large and open in character, providing flexibility and good natural surveillance.

The design consists of two interlocking structures: the first is a generous, lofty gathering space; the second, on two floors, contains customs controls, offices and ticketing facilities. Both structures are supported by curved steel ribs clad in silver profiled aluminium, giving the terminal a sleek elegance alongside the massive cruise liners that keep it company. Transfer to other modes of transport occurs on an adjacent road area with coach, bus and taxi lanes. Since the terminal has translucent walls and glazed gables, there is good visibility to waiting ships and buses and out over Southampton Water. Passengers are able to see their destination with the routes clearly expressed and bathed in light.

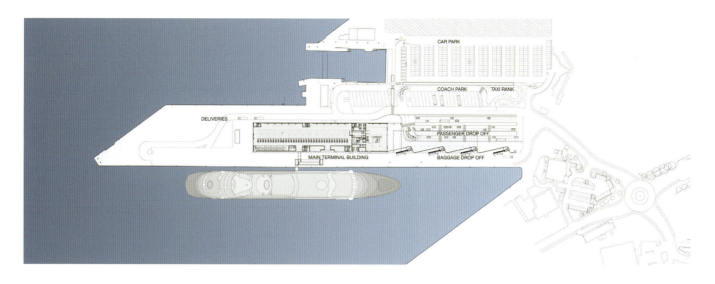

OCEAN CRUISE TERMINAL
SOUTHAMPTON
SITE PLAN

N

THE MANSER PRACTICE ARCHITECTS

7.8 Interior of waiting area at Ocean Terminal, Southampton.
(Photo: courtesy of Morley von Sternberg)

One important requirement of the design was ease of maintenance and protection in the event of a fire. The building uses fire-resistant glazing (supplied by Schucco Jansen), which allows for high levels of daylight penetration while preventing the transfer of radiant heat. Surfaces and details are kept simple to assist cleaning by machine and, where needed, low barriers are provided to prevent baggage trolleys or wheeled suitcases from damaging finishes. This is particularly important where glazing is taken through the roof in the large concourse.

Two New York Ferry Interchanges: Staten Island and St George

The new Staten Island Ferry Terminal, designed by Venturi, Scott Brown and Associates and Anderson/Schwartz Architects, serves around 40,000 commuters and tourists a day (Jones, 2006: 190). Known frequently today as the Whitehall Ferry Terminal, the building acts as an interchange between ferry, subway, bus and taxi services. The interchange is a large inclined glass box with rooftop viewing platforms and promenade decks facing across the water. These give popular views to the Statue of Liberty off shore and to ships plying the Manhattan waterfront.

There is a large glazed waiting and ticketing area, which is built on a raised deck above the harbour-side road network. Passengers are taken across the two-level road via the elevated deck, which serves at its outer face the access bridges to waiting ferries. Hence, the road that previously was a major barrier to movement is now subservient to ferry passengers. It is a good example of the benefit of 'feet before wheels'. The inner face (or landside) of the glazed waiting hall gives access to bus services at ground level and subway ones beneath. As such, the new terminal acts effectively as an interchange in both operational and aesthetic terms. The building is robust in detail with much shuttered concrete, steel, aluminium and glass, yet the attention to harbour-side viewing and promenade allows for the incorporation of large volumes with much transparency and natural light. This encourages the character of the facility to respond positively to the special qualities of the Manhattan site.

Across the water, the St George Ferry Terminal is part of wider improvements to New York's public transport system, which include new or upgraded subway, ferry and bus services. Similar in size to the Whitehall Ferry Terminal, the project involves the restoration of the historic former terminal facilities and their extension to cater for over a million passenger movements per year (Jones, 2006: 182). As with the earlier example, extensive viewing areas have been provided to open up the spectacle of the waterside and Manhattan views to tourists and commuters alike.

The building is an interchange between ferry, tram and bus services. This important function is signalled by an arch of photovoltaic panels that rises above the building and provides some of the electricity needs for lighting, cooling and operational equipment. In this, it is similar to the example of Vauxhall Cross Bus Interchange in London (see page 123) with its angled arms of solar panels. However, the main interest of the St George Ferry Terminal lies in the ability to open up the waterfront, to reorganise movement patterns between transport modes, and to act as a catalyst for wider urban regeneration. The concept is based upon a large harbourside landscaped park with a public promenade overlooking the water. Within the park, a number of new and refurbished older transport buildings are located – the ferry terminal, bus station and tram stop. The geometry of the facilities reflects the dimensions of the different transport systems and their weaving together within the urban park. The arrangement also reconciles two conditions: the urban grid of the city and the natural patterns of the water.

Designed by HOK, the ferry terminal was one of the first in the USA to be LEED certified, making it an object lesson in sustainable practices. These include a green living roof, capture of rainwater for irrigation, photovoltaic panels, maximisation of natural light and cross-ventilation and, in order to enhance local biodiversity, the creation of oyster beds. The latter enhance water quality, attract birds and signal a respect for nature at this busy ferry terminal.

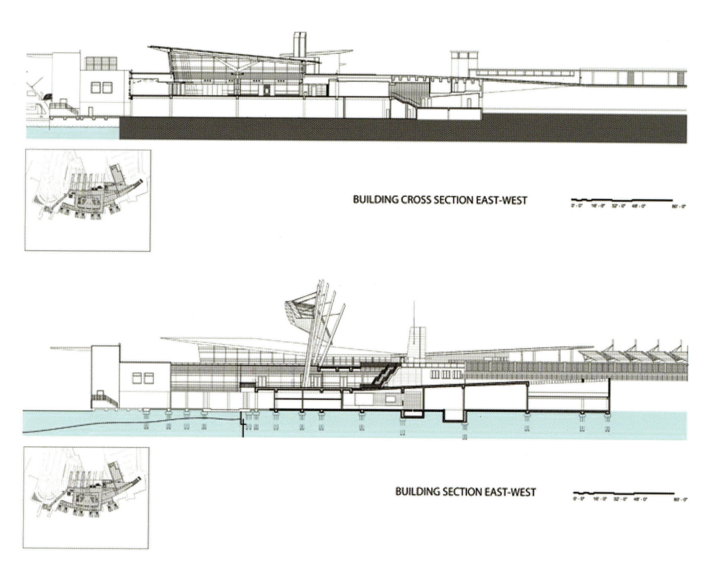

BUILDING CROSS SECTION EAST-WEST

BUILDING SECTION EAST-WEST

7.9 St George Ferry Terminal, New York, designed by HOK. (Photo: courtesy of Adrian Wilson)

7.10 Sections, St George Ferry Terminal, New York, designed by HOK. (Courtesy of HOK)

7.11 Viewing deck, St George Ferry Terminal, New York, designed by HOK. (Photo: courtesy of Adrian Wilson)

Table 7.1 Ferry interchanges

Name	Type	Architect	Key features
Yokohama Ferry Terminal, Japan	Ferry/cruise liner/bus	Foreign Office Architects	• Landscaped roof for public promenade • Wavy shapes
DFDS Terminal, Copenhagen	Ferry/bus	3X Nielson	• Sleek container-like shape • Landmark at night
Salerno Ferry Terminal, Italy (project)	Ferry/cruise liner	Zaha Hadid	• Sculptural shape • Inclined concourse
St George Ferry Terminal, New York	Ferry/bus/metro	HOK	• LEED certified • Many green features • Exploits view
Staten Island Ferry Terminal, New York	Ferry/bus/metro	Anderson/Schwartz with Venturi, Scott Brown	• Upgrading historic terminal • Waterside cafés and viewing decks
Ocean Terminal, Southampton	Cruise/bus/taxi	The Manser Practice with Stride Treglown	• Low-cost transport gateway • Cruise ship interchange facilities

Source: Will Jones (2006).

Airport interchanges

Airport interchanges are always dominated by the airport element. The scale of plane operations, the necessary safety and security demands of air travel, and historic growth patterns of typical airports conspire to reduce opportunities for genuine inter-modality. As a result, the airport interchange is usually a large airport with rail and bus links added, usually later, or placed around the periphery of passenger terminals. However, there are exceptions such as Stansted Airport in the UK, where the train to plane journey was conceived at the outset and resulted in well-integrated interchange facilities. More typically, the original airport has grown into a transport interchange as a result of passenger pressure and enlightened government intervention. Two such examples are Kastrup Airport, with its integral high-speed train station serving both Copenhagen and southern Sweden as well as its new metro to the city centre, and Frankfurt Airport, with its new station served by ICE trains.

One particular problem with airport interchanges is how to accommodate two or more quite different infrastructure arrange-ments. Planes, trains and buses have distinct spatial and engineering demands and reconciling their geometries into a single structure is particularly difficult. Two solutions are commonly adopted. The first is to exploit the cross-section in order to connect the transport systems at different levels. Usually, the train is taken underground, allowing the planes to occupy the surface, and when metro systems are employed (as at San Francisco) they arrive at the airport at high level. Passengers have, therefore, to navigate between transport decks, which, with large bags and baggage trolleys, can pose difficulties. The other strategy is to place the inter-change facilities around the edge of airport terminals and connect them by a light rail system, shuttle bus or moving pavements (as at Lyon Satolas Airport and Chicago O'Hare). Often, the two strategies are combined to produce hybrid interchange facilities, as at Schiphol Airport in Holland and Charles de Gaulle Airport in Paris.

In Europe, the development of the high-speed train network has had a big impact on the continent's airports. These are now being converted from airports into modern efficient interchanges by their connection to TGV or ICE rail services. Although this is true of all airports, it is particularly the case with hub airports, where high-speed rail connections allow for surface completion of journeys at a regional level (up to 150 kilometres).

In designing airport interchanges, or when adding new transport services to existing airports, the following points should be considered:

- Can the passenger see the connecting services from the air, on their arrival at ground or after clearing customs controls? It is important that visual links are provided since many travellers will be tired after often lengthy journeys and may not be able to read the direction signs.
- Is there integrated ticketing and integrated timetabling? Can the passenger see the times of trains or buses while still in the terminal building and can they buy tickets there? Ideally, the tickets will be interchangeable between modes of travel.
- Is there adequate provision for carrying bulky baggage? This applies particularly to the escalators, lifts and stairs. Wide passageways are needed, not just to ensure visual linkage, but to accommodate passengers and their baggage.
- Do the linking services cater for all types of passenger (disabled and elderly) and are the transport modes designed for excess baggage?

As with all journeys, movement is in two directions. Hence, knowing where you are going and where you have come from is important. Avoiding disorientation means using design in both plan and section to steer the passenger towards the journey objective. It also entails using effects of light, material or art to punctuate the journey, especially at critical points such as crossings or changes of angle or level. Architecture here is a matter of designing for cognition rather than simply satisfying functional demands.

Stansted Airport

Stansted Airport, designed by Foster + Partners in 1987, was one of the first airports in the UK to be conceived as a transport interchange. The inclusion of a direct rail link to central London in the brief issued to the architects was a major departure from airport architecture at the time. Although the train provision is fairly modest compared to practice today, particularly in mainland Europe, the benefit to passengers was considerable.

8.1 O'Hare Airport, Chicago. Internal link from train to plane. Notice the use of art to enliven the travelator journey. (Photo: Brian Edwards)

8.2 Passenger concourse at Frankfurt Airport Station, designed by BRT. It provides an elegant link to high-speed ICE train services. (Photo: Brian Edwards)

Today, the twice-hourly train links to Stansted compensate for its relative distance from London and the geographic position of the airport within the largely agricultural region of East Anglia. However, in spite of good rail and motorway links, local planning policy is currently preventing the airport from expanding into a full interchange and magnet for sustainable development.

The train station is positioned below the main airport concourse levels, rather than alongside them, as in the manner of Frankfurt or Manchester Airports. This has the advantage of proximity, but the disadvantages of vertical movement for passengers with large amounts of baggage and also lack of daylight. Foster's concept sketch plan showed the relationship between plane, train, bus and car, which the terminal successfully integrates. Trains enter at low level (and hence are below the runways and taxi areas) and terminate beneath the concourse on its landside. A shaft of light is taken down into the platform areas, which otherwise are artificially lit. Large columns necessary to support the terminal above articulate the platforms, giving them a heroic scale. Ramps, escalators and lifts take passengers up to the check-in concourse and past the ticket area.

The rectangular discipline of the airport terminal absorbs the sub-structures of the railway element. The language of columns, diagonal rooflights and perimeter glazing, which is such a feature of Stansted, is not interrupted by the inclusion of the station. This nestles unobtrusively beneath the terminal building. It is complemented on the airside by another rail system, this time the light rail trams that transport passengers to the satellite piers for departure. In many ways, Stansted is a model of well-integrated transport from plane to heavy rail, light rail and bus. The major problem today is that the popularity of the airport and the infilling of concourse areas with shops, cafés and large advertising banners makes finding the various transport services difficult at times.

Frankfurt Airport

The construction of the new station in 1999 at Frankfurt Airport has not only allowed ICE trains to serve Germany's busiest airport, but has converted Frankfurt Airport into a highly efficient transport interchange. Although buses and regional trains previously served the airport, the high-speed station allows Frankfurt to be a hub for rapid surface travel as well as in the air.

Since high-speed trains are particularly long and demanding in terms of track engineering, they cannot easily be accommodated under or close to airport terminal buildings. The arrangement at Frankfurt was to build the new high-speed rail station beyond the existing regional train facilities and to link them via a 600 metre-high level walkway beneath which roads and airport service vehicles pass. In simple terms, the rail links run at right angles to the axis of the airport and are located between the terminal buildings and an existing motorway.

The new station is 400 metres long and cut into the ground. Light enters via a dramatic oval roof light over the platforms and this provides a point of orientation at the centre of the complex. Above the station a hotel 600 metres long takes advantage of the air rights and helps landmark the new facilities within the wider airport landscape. Since the hotel 'floats' above the station, light enters the caverns below

8.3 Platform at Frankfurt Airport Station. (Photo: Brian Edwards)

8.4 Concourse at Kastrup Airport, Copenhagen. The train ticket office is to the right with trains passing immediately underneath. (Photo: Brian Edwards)

ground. The relationships between the station, hotel and airport are clearly articulated by architects Bothe Richter Teherani (BRT), using an expressive technology of glass, steel and concrete. One feature of the design is the use of air locks in the train tunnel to allow comfortable conditions to exist at platform level.

Schiphol Airport

Schiphol Airport is the hub for KLM and is served by many other carriers. The location of Schiphol and its extensive rail links allow the airport to serve all of Holland's major cities and those in adjoining Belgium, Luxembourg and Germany. The airport has a compact, shallow, V-shaped centre with finger piers leading to boarding gates. Trains are taken into the airport directly beneath the main central concourse. Passengers merely take the escalators up from platform level to the airport concourse, where the train ticket offices are also located.

The airport station is served by local and regional trains as well as the ICE International rail network. One weakness of the arrangement is the lack of daylight in the station. Being beneath the airport it does not have access to daylight or sunlight, which

is a major drawback of this type of arrangement. Also, since some of the trains are diesel fuelled, there is the added problem of air quality within the subterranean station. Added to this, it is difficult to perceive the presence of the station from within the terminal buildings, since, unlike at Stansted, it is not opened up to view.

Schiphol is also well served by buses, which stop at the airport entrance within the embrace of the 'V'. Hence, movement between rail, bus and plane is relatively straightforward, although the lack of visual communication may disadvantage those passengers unfamiliar with the airport. The airport offers good facilities for bike users, allowing cycling to play its full part in transportation.

Kastrup Airport, Copenhagen

The new railway station at Kastrup Airport is built beneath an arrow-shaped extension to Terminal 3. In effect, it is a

8.5 Train at Kastrup Airport Station, Copenhagen, with view of airport check-in terminal behind. (Photo: Brian Edwards)

8.6 Bus stop at Kastrup Airport, Copenhagen, with check-in concourse behind. (Photo: Brian Edwards)

8.7 Sheltered bike storage at Kastrup Airport, Copenhagen. (Photo: Brian Edwards)

on a shallow curve, with the platforms protected from the elements by elegant glass roofs. A series of large openings allow views up to the concourse and help orientate passengers at their arrival at the station.

Besides the rail and metro links, Kastrup provides bus stops for local services adjacent to the terminals and also extensive cycle storage. Kastrup is linked to central Copenhagen by cycle lanes and these are proving increasingly popular, particularly in the summer. In many ways, Kastrup is a model of inter-modal connection to an airport. Trains run every 10 minutes to the city centre and every 20 to Malmö in Sweden using the new Øresund Bridge. In addition, driverless metro trains run every 5 minutes during airport operational times of 5 a.m. to midnight. What makes the arrangement different from elsewhere is the attention paid to ensuring visibility of the connecting services and the adequate space on trains and metro facilities for baggage. Also, all the connecting rail and metro stations are served by lifts, making travel available to all.

Inchon Airport Interchange, Seoul, South Korea

Designed by The Farrell Partnership, the transportation centre at Seoul International Airport is a model of interchange efficiency. The building works as a reception point for air passengers arriving by rail, car or bus and, in the opposite direction, for those disembarking from flights. The huge structure has the task of receiving and directing large flows of people, providing travel information and ticketing, and giving passengers a meeting point free of immigration controls. It does this within an envelope of high environmental standards, where particular attention is paid to natural light, solar shading and cross-ventilation. One consequence of this is the way in which the interchange provides route clarity and directional orientation because of the inclusion of a sense of the external environment within the internal volumes.

The transportation centre at Inchon is shaped like a fan angled towards the runways. This allows the main functional patterns of the airport to impact upon the interchange – adding again to the grasp of operational patterns. The passenger

bridging structure between Terminals 2 and 3 that extends out to form a link to the high-speed rail line, which crosses at right angles at low level and to the metro services at high level. The geometries are straightforward: the trains pass beneath the V-shaped concourse at about its midpoint and the metro services extend the concourse at its tip. The wide part of the 'V' splits passengers into those moving to Terminals 2 and 3.

The interchange element of Kastrup is a long, delta-winged, V-shaped structure with a central rooflight above designed by Vilhelm Lauritzen Architects. It is glazed on all sides, allowing the decks of the platforms to be viewed below and the metro services beyond. The space is surrounded at high level by shops and cafés, thereby bringing life to space that is occupied in a transitory fashion. The ticket office for the trains occupies an island within the concourse and lifts, ramps and stairs run off in an orderly fashion. A pair of tall, steel columns provide a spine of support for the roof and help define a central thoroughfare through the space. The station at low level enters

Table 8.1 Airport interchanges

Name	Type	Architect	Characteristics
Stansted, England	Plane/train/bus	Foster + Partners	• Station beneath terminal • Natural light and ventilation
Schiphol, Amsterdam	Plane/train/bus/cycle	Various	• Concentrated interchange • Full inter-modality
Kastrup, Copenhagen	Plane/train/metro/bus	Various	• Light used for navigation • Integrated into single 'V'-shaped building
Frankfurt, Germany	Plane/train	BRT	• Well-lit concourse • Visible transport links • Landmark
Manchester, England	Plane/train/bus	Various, including Austin-Smith: Lord	• Distinctive landmark shape • Travelators to different terminals • Air rights exploited
Inchon, Seoul, South Korea	Plane/train/bus/ park and ride	Farrell Partnership	• Single interchange for large airport • Shape based upon fan • High environmental standards

moves up from the four-track railway station into a large triple-height concourse, through which runs an elevated light railway. The latter leads to the various terminal buildings and, in the opposite direction, to the bus points and car parks. Planning in three-dimensional space is the key to success at Inchon, making the transportation centre a possible model for urban applications. The shape in section reflects not only the patterns of movement within, but also the environmental profiles necessary to reduce energy consumption for cooling and lighting. The use of the fan as a design metaphor alludes to both cultural and environmental traditions in the region, thereby adding richness in a social and political sense.

PART 4

Conclusions

Thoughts and future issues

This final chapter speculates on some of the themes discussed earlier. It seeks to raise issues that may be of interest to those who design, manage and fund public transport provision. The focus is on the transport interchange, but questions are also raised about associated topics, such as sustainable development and social inclusion. The aim is to summarise the argument, to highlight areas that have been neglected in related transport and sustainability scholarship, and to point the way forward.

The conclusions are presented as speculations. The topic of the transport interchange is fast evolving, with investment shifting significantly from roads to rail and from private mobility provision to public. Nowhere is this more evident than in China, where the interchange has found new forms of expression. The scale and complexity of transport provision affects the physical, economic and social structure of cities. Hence, one important conclusion is that of 'placemaking' rather than just building-making. Another is the key relationship between technology, space and architecture, and the importance of collaboration at the start of transport projects between architects and engineers.

There are eight speculations in the conclusions. These are a mixture of thoughts and guidance for the future shaping of this important transport building type. Each topic follows roughly in chronological order their position within the flow of the book. However, certain themes are so embedded that they appear first, while others are structured because of the inter-relationships between the topics.

The search for sustainable development

Although the interchange is currently evolving into a new recognisable transport type, the path and pace of evolution vary between continents and are being differentially shaped by forces beyond the sphere of transportation provision. One major external force is that of sustainable development and the associated concerns over global carbon emissions. Moving people by public transport is far less energy intensive than by car and, although the relative fuel efficiency of different modes of public transport varies, mobility by train, bus, ferry, and even air, results in lower contributions to climate change. So, one prime incentive to invest in public transportation is that of environmental sustainability. However, sustainable development touches also upon issues of

social and economic sustainability. Hence, unlike the individual rail or bus station, the transport interchange provides the opportunity to integrate social and welfare provision within a framework of economic and business potential. One key conclusion, therefore, is that the rise in the concept of the interchange is a direct response to the political and cultural shift towards the attainment of sustainable development within its three paradigms: environmental, social and economic. As Sir Terry Farrell notes, 'successful economies need bold long-term investment in railway infrastructure' (www.e-architect.co.uk/beijing/beijing_south_railway_station) and the scale of this in countries such as China, France and Germany is one of the drivers towards both the interchange and sustainable development.

However, to achieve its full potential in contributing towards sustainable development another trend is apparent. This concerns the concept of ribbons that join the transport network to the wider community. Attention to feeder routes and integrated services, both on foot or by bicycle, and by bus and

light rail, helps the interchange to better meet the mobility needs of the twenty-first century. In doing so, it brings urban planning to the fore as a discipline that connects the interests of land use and transportation. So, whereas the interchange is a building or specific place of transport interconnection, its ramifications on city form, urban regeneration and social inclusion are wide. This has been overlooked in the past. Too often the station or interchange was funded, designed and engineered as a problem of transportation, while in reality it is a problem of sustainable development.

Social inclusion

The transport interchange is a key instrument in achieving social inclusion. Whereas the private car promotes the separation of classes and urban dispersal, the interchange encourages social and cultural integration. By mixing together people, land use functions and movement webs, the transport interchange

9.1 The new square outside Sheffield Station provides a generous space for meeting and connecting to bus and taxi services. Converting former railway stations into interchanges is one of the challenges ahead. (Photo: Brian Edwards)

adds to social cohesion and community well-being. This is particularly important in the inner cities, where many large-scale transport facilities are located. Both directly and indirectly, the interchange can become the location for social and physical regeneration. Public transport encourages people mixing, cultural dialogue and the maintenance of civic values.

As the human population ages in developed western countries such as the UK, the role of public transport in the quality of life increases. For many people, mobility is restricted by physical and economic limitations. Expanding access to facilities for the elderly and the poor is an important dimension of the interchange. However, to achieve this, attention needs to be paid to those who are less able and to those who become easily confused. Hence, the transport interchange needs to provide clear routes, avoid unnecessary changes in level and direction, be well lit but without glare, have sitting areas for resting, and have adequate toilets and supporting social facilities. There should also be a single ticketing policy, thereby allowing for easy transfer between transport modes, integrated travel information and well-displayed clocks. To achieve these social goals, the commercial frontiers should be kept at bay, with retail operating within defined zones that are secondary to the movement and transfer areas.

Types and typologies

The rise of the interchange as a concept is a reflection of the natural evolution of building types. As society becomes more complex in its movement patterns, and in the distribution of economic, social and cultural services, there is a corresponding swing from singular building typologies to hybrid ones. Over the decades, singular types such as railway stations have evolved into complex interchanges – the latter growing into a recognisable typology. The same is true of airports, where the addition of new rail and metro connections has turned once-isolated airports into multi-functioning transport hubs. Hybridisation means that many modern interchanges carry the genetic inheritance of past types. These genes are mixed to aid the process of evolution. So, we have within the cannon of interchange types those that are bus interchanges, others that are rail interchanges etc. Generally speaking, the older

the interchange, the more complex the transport connections. Also, older interchanges perform a wider range of social, cultural and economic services than modern ones. A lesson from urban history is that space must be provided for the growth of the social and commercial activities that inevitably occur within or adjacent to mature interchanges.

As types mutate and new typologies evolve, new design and engineering solutions are needed. Looking at recent interchange construction, it is evident that experimentation is richest in Asia. The focus of architectural invention in transport is found particularly in China, but also to a lesser extent in Japan and South Korea. In the emerging economies there is less inherited infrastructure that needs adapting and certainly fewer listed buildings, which can inhibit invention in Europe and the USA. But as a consequence of the compromises found in provision in Europe, the interchange is often better integrated within the city.

Within the typological distinctions discussed above, there are examples of large interchanges and small ones. Often there is greater clarity of movement in smaller interchanges and their connection to urban centres is less congested. Suburban interchanges are a particular type, particularly those that join bus and rail services. As interchanges grow in size, so too do the problems. Big urban interchanges may connect four or more modes of transport, often operating at three or more levels. Frequently, too, there are inherited structures that need to be preserved. The pressure to resolve weaknesses from past arrangements while meeting the aspirations of a demanding public are enormous. With public transport use in Europe increasing at about 6 per cent per year, the designer needs to search for new typological and spatial orders to solve the problems.

The interchange mirrors the growing hybridisation and complexity of modern urban life. As such, it is a place of many stakeholders, many compromises and many conflicting agendas. To reach consensus there needs to be leadership and humility. A question raised in the interviews conducted for this book concerns who is best able to offer the leadership necessary to provide the fusion of urban planning, design, engineering and procurement skills. Traditionally, it has been the engineer representing the main infrastructure company involved, but today the architect, and increasingly in Asia the urban designer, is taking the lead. Design skills are essential if

9.2 Technology has been a key driver in upgrading transport and its supporting buildings. This example is Flintholm Interchange north of Copenhagen, which won the Brunel Award for architecture in 2005. (Photo: Brian Edwards)

the interests of passengers are to be protected under the impact of often considerable cost and technical difficulties. However, too often the designer's role is limited to that of a few conceptual statements, sketches and diagrams that win a commission. The implementation then flows down traditional lines with scant regard paid to the quality of detailing or material selection. This trend is not good for lasting sustainable development, particularly when full life cycle costing is considered.

Signally sustainable infrastructure

One characteristic of many recent interchanges is their use of ecological approaches to design. This has included working within natural systems for waste recycling, using locally sourced construction materials, exploiting natural light and ventilation, using passive solar gain for winter heating and summer ventilation, and incorporating new energy technologies such as PV cells and geothermal energy. Some recent transport buildings have used their height and large surface area to exploit ever more efficient photovoltaic technologies, thereby producing electricity to use in the building while also sending an important message to the wider community. Others have used ground-source heat for warming, exploiting the geothermal energy associated with tunnelling. Within the architectural frame, green materials and green energy technologies project an important message to users of interchanges, who can count into millions over a typical year.

With many transport interchanges, the energy problem is that of cooling rather than heating. Here, again, many innovations can be found. Ground sources used for winter warming (using heat pumps) can be utilised for summer cooling. As interchanges have grown in scale and civic importance, their role in applying new energy technologies has gained momentum. With increasing scale come the inter-relationships necessary to deliver green initiatives. Waste from one area can become the raw materials for another, big buildings (such as interchanges) can support in energy terms smaller ones round about, and water capture can be utilised for evaporative cooling and local plant irrigation. Also, the economic power of transport interchanges can provide the finance necessary to regenerate redundant buildings in the vicinity, in the process creating

jobs and helping establish local industries. Sustainable infrastructure is more than applying photovoltaic panels to the roof (although this sends an important message to the community); it is about adopting an integrated ecological view. The lesson from many of the examples in this book is the value attached to an integrated approach to sustainable infrastructure from the urban to the building scale.

Technological innovation in travel

If broad social drivers for sustainability have been behind the rise of the interchange, another important influence has been that of technological innovation. Developments in high-speed rail, in new bus traction, and in the engineering and efficiency of metro systems, ferry boats and aircraft, have all contributed towards the emergence of new interchanges. One key area has been the cementing together of existing systems with a number of technical innovations in bus and metro technologies. Automatic signalling and the use of driverless trains, the application of bio-fuels and new combustion technologies, electric vehicles, bus guide lanes and flexible bus formations have helped in reconnecting existing underutilised transport facilities. This in turn has increased user levels, which has acted as a motor for fresh investment. From single provision has risen the interchange, due largely to better use of road and air space around inherited transport infrastructure for a diversity of connecting transport provision. In the wake of technological innovation and a consequent diminishing use of cars and polluting buses for urban transport, there has been an improvement in air quality. This has made the interchange a place rather than merely a connecting building.

One of the major technological innovations has been in high-speed rail. This has had great influence in Europe and Asia. Many cities are now joined not by air or slow regional trains or buses, but by fast, comfortable trains. TGV, ICE and their Chinese and Japanese equivalents have revolutionised national travel and even impacted on international journey choices. Connecting high-speed networks to existing stations and termini has led to considerable enhancement of travel facilities in many cities. However, to take advantage of such travel there needs to be connection by other feeder routes – metro,

9.3 By combining small retail units in this bus and tram shelter in Vienna, the sense of social sustainability is enhanced. (Photo: Brian Edwards)

bus, regional train, and even air. The pattern has been one of upgrading facilities, functionally and aesthetically, to cater for high-speed rail provision.

There have been innovations, too, in the mode of moving people at interchanges. Many facilities associated with airports, such as travelators, escalators, spacious glazed lifts and electrically powered people movers, are now found in bus and railway interchanges. In an attempt to cater for all levels of mobility, the interchange has adopted high standards of inter-connecting provision. This aids both the commuter who rushes through the spaces provided and the tourist or visitor who is there for the experience. So the airport experience with both its positive and negative qualities is now impacting upon the urban interchange.

Technological innovation has changed both the mode of travel and the associated transport facilities. Here, there are different forces at work in different regions of the world. As discussed, the development of high-speed rail has had a big influence in Europe and Asia, but in South America new bus technologies have driven the emergence of small local inter-changes. New metro provision and hybrid bus/metro systems are a major force for change in the USA, often funded by state or city authorities. These occur in suburban locations where fuel poverty is often acute. In the USA also, the pattern is one of joining together the fragments of systems that were built in isolation. Here, the interchange is a connector and land-mark utilising new low-carbon technologies to signal a change to a post-fossil fuel economy. So, whereas Europe, Japan and China provide examples of big transnational interchanges, more modest but still important interchanges are an increasing feature of new American urbanism. The latter trend is evident in Africa and countries such as Thailand and Vietnam, where the interchange may involve small local buses, rickshaws, motorbikes and bicycles.

An important and often overlooked connector to public transport is the bicycle. Cycle technologies have also seen remarkable innovation, widening the appeal of the cycle, particularly to the elderly. Hybrid pedal and battery-driven bicycles, tricycles and box formats are now quite common, particularly in cycle-friendly towns and cities such as Seattle, Davis in California, Amsterdam and Copenhagen. What is evident, however, is the poor provision for cycle storage at many public transport facilities. The interchange is often the point of transfer between bicycle and bus or train, yet too little attention is paid to the bicycle and the cyclist.

Place or space

Typical building briefs for transport buildings give floor areas needed for different functions. These programmatic criteria tend to limit the plan to meeting merely functional demands. Hence, space is provided in a utilitarian sense at the expense of place. One trend discernible in recent projects is the striving for a sense of place in the provision of transport buildings. This is particularly marked in larger and more complex interchanges, where place awareness tends to reinforce the perception of functional zones and key routes. Place is also an important point of reference in giving cultural identity to modern transport buildings.

The creation of place requires considerable design input. Places are not just big spaces, as many architects and urban designers know. However, not all commissioning bodies and civic clients recognise the distinction, and often budget guidelines have a negative impact on place quality. Yet place is what makes space special – it is the ingredient that ensures survival over time. At interchanges there need to be both space (for functions) and place (for use recognition and pleasure). Interchanges should lead to a state of happiness, not confusion and depression. The concept of happiness in architecture is difficult, yet many transport interchanges are alienating and unattractive in spite of generous space provision. Key qualities of place are light, clarity, celebration, and social and cultural activity. Places are normally enclosed, animated by human activity, not overcrowded or cluttered, and landmarked or given dignity by façades and surfaces of quality.

At transport interchanges, places occur inside and out. The central concourse represents a place that is the focal point for transfer and information activity. It is normally fairly spacious, but transcends physical dimensions by the quality of design and the level of service provided. As a transport building there will be much activity, but place awareness resides in the quality the space has, not its actual dimensions. Such quality can be expressed in the proportions (the height to width ratio), the

9.4 Holyhead ferry, rail and bus interchange, designed by The Manser Practice. (Photo: courtesy of Chris Gascoigne)

9.5 Sun-shaded car park at Atocha Station, Madrid, designed by Rafael Moneo. The interchange has often to cater for park and ride facilities, which here are placed on the roof to shade the concourse spaces below. (Photo: Brian Edwards)

clarity of routes through the space, the quality of the materials employed, and the colours, lighting, detailing and furniture. Place is subjective (space is not), hence there will be different perceptions held of the same area by different users. To a degree, space becomes place over time due to human intervention, but conversely the initial creation of place quality can be eroded by poor management of facilities.

Place occurs inside and outside the interchange. One characteristic of the interchange, particularly on mainland Europe, is the presence of generous external gathering space. This usually links into the wider network of routes joining the interchange to the city. Hence, the entrance to the travel facilities has already given the passenger the opportunity to prepare for the journey or, if travelling in the opposite direction, to orientate to the city. The forming of place gives the opportunity to restructure the hinterland of interchanges. Many recent examples show the power of transport buildings to transform neighbourhoods into districts with character and quality.

Weaving and flow

One important feature of the interchange is the ability to move big volumes of people involved in complex interactions with the spatial dynamics of transport systems. Usually, there are three types of flow – that of people, that of the facilities and that of transport. However, there may be other flows such as the movement of people through transport buildings who have no intention of travelling. There are also usually vehicles used to help people through the flow, such as lifts and buggies. So the dynamics of transfer, particularly the patterns of flow, are

an important force in shaping the architecture of interchanges. This has found expression in the weaving together of architectural forms and the creation of undulating planes that mirror the processes of interconnection. Since transfer is three-dimensional, such spaces and built elements have a sculptural quality. This in turn has, either deliberately or accidentally, led to the emergence of buildings with landmark qualities on the one hand and landscape ones on the other.

Connection is also a visual phenomenon. The ability to see the point of arrival or mode of departure has resulted in interchanges that are highly transparent. The use of visibility has replaced that of signs in aiding the passenger. However, such visibility is often undermined by later insertions into what were intended to be open visual corridors. Here, intrusion by retail elements and advertising is a major problem in terms of passenger flow. Open concourses have been replaced by shopping malls, with the result that ticket facilities and travel information boards are obscured. Sometimes it seems that flow and information are deliberately impeded by commercial interests. Too often at interchanges the sense of service is secondary to that of profit. To overcome this there needs to be stronger continuity between the architectural vision and the subsequent reality of use.

Weaving and flow are also major factors for the transport services found at the interchange. One feature discernible in larger interchanges is the extent of infrastructure compromise necessary to join together a transport system that may originally have been conceived as separate. The creation of the transport interchange frequently involves making better connections between existing systems, with perhaps a new metro line acting as the cementing agent. Normally, the interconnection is via

small flexible transport systems, rather than large ones. Hence, bus and metro systems join regional and high-speed rail or airport services. The inflexibility of heavy rail means that light rail and rubber tyres are the gel that carries the connecting load. Putting the engineering systems into context, it means that there is a hierarchy of flow from the passenger upwards, which is feet (humans) to narrow wheels (bicycles) to wide wheels (buses) to hard wheels (light rail then heavy rail) and to large wheels (planes). This simplistic summary is a reminder of the importance of the human dimension.

Leadership and compromise

The interchange is a building where many stakeholders come together. Often, the complex interests undermine the quality and speed of decision making. Unlike many other building projects, the transport interchange involves a multitude of agencies, funding bodies, infrastructure and community interests. At a typical project design meeting there may be 30 or more clients represented. The question of leadership is important, as is the nature of the brief or programme, and the availability of funding to meet a broad range of social, transportation and economic objectives. Leadership involves negotiating compromise in the interests of the project. Leadership is also a question of vision, design aspiration and commitment. Whether these are best delivered by the architect, engineer or project manager is a moot

point. Of the projects described earlier, many have been led by engineering consortia, others by architects, and some by the client's own project managers.

Since many interchanges are the result of upgrading existing facilities, some of which have existed for decades, the compromises today are compounded by the compromises made in the past. Hence, much effort is put into removing unsatisfactory elements from the past in an attempt to clarify arrangements for the future. Entrenched positions can impede the achievement of a high-quality transport environment. This can occur with regard to heritage elements on the one hand and informal relationships with retailers on the other. Strong leadership is needed to overcome these barriers.

One trend is towards the emergence of named architects for projects. A number of readily recognised international architects now compete for some of the world's prestigious transport projects. They have the experience and authority to carry their visions into reality. However, the knowledge base for transport projects, particularly interchanges, is high. This results in the use of large consortia of architects, engineers and contractors tendering for work. Leadership has shifted from the architect to the engineer and now to the project manager and contractor. The change mirrors the importance of controlling cost and quality as perceived by government. However, in the process it is important that the design vision is safeguarded, and also the attention to detail upon which the question of quality for the consumer often resides.

Bibliography

ArchIdea (2009) 'Interview, Sir Nicholas Grimshaw', *ArchIdea*, 39.

Auge, M. (1995) *Non-places: Introduction to an Anthropology of Super-modernity*, London: Verso Books.

Blow, C. (2005) *Transport Terminals and Modal Interchanges: Planning and Design*, Oxford: Architectural Press.

Browne, B. (ed.) (2008) *Jean-Marie Duthilleul and Etienne Tricaud: AREP*, Mulgrave, Vic.: Images Publishing.

Chapman, R. (2009) 'Chatswood Transport Interchange', *Architecture Australia*, 98(3) (May/June).

Collis, H. (2003) *Transport, Engineering and Architecture*, Oxford: Architectural Press.

Curtis, W.J.R. (2008) 'The Stirling Prize is open to the vagaries of taste', *The Architects' Journal*, 24 July.

Dobrovolsky, L. (2009) Interview with author, 1 December. Leszeck Dobrovolsky is a Director of Arup.

Edwards, B. (1997) *The Modern Station: New Approaches to Railway Architecture*, London: Spon Press.

Edwards, B. (2003) *Modern Airport Architecture*, London: Spon Press.

Foster, N. (2008) 'Foreword', in Niedenthal, C. (ed.) *Stations in Germany*, Berlin: Jovis Verlag, pp. 9–11.

Fraser, K., Pugh, R. and Wong, D. (2005) 'Vauxhall Cross bus station, London', *Arup Journal*, 3: 3–6.

Freudendal-Pedersen, M. (2009) Interview with author, 8 December. See also Freudendal-Pedersen, M. (2009) *Mobility in Daily Life: Between Freedom and UnFreedom*, Farnham: Ashgate.

Grimshaw, N. (2009) Interview with author, 24 June.

Harbour, I. (2006) 'Foreword', in Jones, W. (ed.) *New Transport Architecture*, London: Mitchell Beazley, p. 6.

Hegger, M., Fuchs, M., Stark, T. and Zeumer, M. (2008) *Energy Manual: Sustainable Architecture*, Edition Detail, Basel: Birkhäuser.

Jones, W. (ed.) (2006) *New Transport Architecture*, London: Mitchell Beazley.

Juul-Sørensen, N. (2009a) 'Sustainable transport', lecture given at the Royal Danish Academy of Fine Arts, 23 September.

Juul-Sørensen, N. (2009b) Interview with author, 5 October. Nille Juul-Sørensen is the lead architect at Arup, designing the new Copenhagen Metro.

Kolb, J. (2008) 'On track', *The Architects' Journal*, 3 April: 27–39.

Le Corbusier (1927) *Towards a New Architecture*, Paris: Edition Cres.

Lowenstein, O. (2006) 'Environmental rail: building design', in Jones, W. (ed.) *New Transport Architecture*, London: Mitchell Beazley.

Lynch, K. (1960) *The Image of the City*, Cambridge, MA: MIT Press.

The Manser Practice (2007) Press release given to the author.

Marsay, A. (2008) 'Sustainable transport: a case study of South Africa', *Arup Journal*, 3: 37.

Niedenthal, C. (ed.) (2008) *Stations in Germany*, Berlin: Jovis Verlag.

Nielsen, J. (2009) Interview with author, 30 October.

Pevsner, N. (1976) *A History of Building Types*, Princeton, NJ: Princeton University Press.

Raisbeck, P. (2007) 'Bridging the void', *Architecture Australia*, Jan/Feb: 45–55.

Richards, B. (2001) *Future Transport in Cities*, London: Spon Press.

Ritchie, A. and Thomas, R. (eds) (2009) *Sustainable Urban Design*, 2nd edition, London: Spon Press. See particularly pp. 3–10, 17–19, 29.

Rogers, R. (2001) *Towards an Urban Renaissance*, Urban Task Force Report, London: Urban Task Force.

Rogers, R. (2009) Interview published in *Architecture Today*, 200 (July/August): 34.

Ross, J. (ed.) (2000) *Railway Stations: Planning, Design and Management*, Oxford: Architectural Press.

Schilperoord, P. (2006) *Future Tech: Innovations in Transportation*, London: Black Dog Publishing.

Scott, F. (2003) *InterchangeABLE*, London: Royal College of Art in association with Scott Brownrigg. Available online at www.hhc.rca.ac.uk/resources/publications/CaseStudies/id4218.pdf (accessed 30 July 2010).

Spurr, S. (2009) 'The Underground', *Architecture Australia*, 98(3) (May/June): 55–61.

Steffin, A. (2008) 'World changing', lecture given at the Danish Architecture Centre, 30 October.

Tempelman, E. (1999) *Sustainable Transport and Advanced Materials*, Delft: Eburon.

Thalis, T. (2007) 'Parramatta Transport Interchange', *Architecture Australia*, Jan/Feb: 56–63.

Tolley, R. (ed.) (2003) *Sustainable Transport: Planning for Walking and Cycling in Urban Environments*, Boca Raton, FL: CRC.

Wong, S. (ed.) (2008) *UK>HK: Farrells Placemaking – from London to Hong Kong and Beyond*. Hong Kong: MCCM Creations.

Wortman, M. (ed.) (2005) *Public Transport: On the Move*, New York: Visual Reference Publications.

Journals

Architectural Review
Architecture Australia
Architecture Today
Arkitektur
Detail

Websites

www.cabe.org.uk (Commission for Architecture and the Built Environment)
www.cfit.gov.uk (Commission for Integrated Transport)
www.tfl.gov.uk (Transport for London)

Index